THE ART OF PRECAST CONCRETE

David Bennett

THE ART OF PRECAST CONCRETE
COLOUR TEXTURE EXPRESSION

Birkhäuser – Publishers for Architecture
Basel · Berlin · Boston

We would like to thank the following institutions
who kindly supported this publication:

Aalborg Portland, Aalborg, Denmark
Betongvaruindustrin, Danderyd, Sweden
Bundesverband der Deutschen Zementindustrie e.V., Berlin, Germany
Lafarge, Paris, France
Rakennusteollisuus, Helsinki, Finland
The Concrete Centre, Camberley, England

Graphic design: Alexandra Zöller, Berlin

Parts of "Precast Materials and Methods of Manufacture" are derived from the essay "Cast Reconstructed Stone" written by David Bennett for the publication Christoph Mäckler (ed.), *Material Stone: Constructions and Technologies for Contemporary Architecture*, Basel: Birkhäuser, 2004.

A CIP catalogue record for this book is available from the Library of Congress, Washington D.C., USA

Bibliographic information published by Die Deutsche Bibliothek
Die Deutsche Bibliothek lists this publication in the Deutsche Nationalbibliografie; detailed bibliographic data is available in the internet at http://dnb.ddb.de.

P.O. Box 133, CH-4010 Basel, Switzerland
Part of Springer Science+Business Media
Printed on acid-free paper
produced from chlorine-free pulp. TCF ∞

Printed in Germany
ISBN-13: 978-3-7643-7150-0
ISBN-10: 7643-7150-1

9 8 7 6 5 4 3 2 1

www.birkhauser.ch

CONTENTS

PREFACE

The Brutalist period that followed the Modern Movement era, where cast in place concrete was used to excess, led to decades of mistrust and rejection of its architectural merit as the rust stained, grey-black, pock marked surfaces were laid bare for all to witness. Precast concrete up to that time was largely specified for making reconstructed stone panels, paving slabs and decorative features. Now it was used as the replacement for cast in place concrete in many European countries, ensuring the integrity of surface appearance with off-site manufacture. Precast concrete's popularity grew, the product range increased and many new precast companies started up. Colour, surface texture, light and shade profiling and bas relief effects plus plasticity of form and large panel construction gave architects a design freedom that was not possible with cast in place concrete.

Better material understanding, researches into surface durability, improved standards of manufacture and production continues to position precast concrete as the premier product for surface appearance in Northern Europe but it comes with a warning. It can be prohibitively expensive in some countries and is not always a popular choice.

In today's building markets, reflected by its share of the architectural cladding market, popularity of precast concrete varies dramatically across Europe. In the UK for example it is considered the most expensive, heavy-weight cladding option for a façade. Its market share is less than 2% of the cladding market. In Finland precast concrete takes 33% of the total building market and is the most dominant material for cladding multi-storey residential buildings (97%) because it is the cheapest and most efficient method of construction. Precast market situations in Sweden and Denmark echo the trend in Finland.

In researching the material for this book this startling difference in market share became all too transparent. Market share is higher where the price of precast cladding panels is low or competitive with alternatives – that is obvious. What is not are the reasons for these big differences.

In Finland to keep precast prices competitive architects and specifiers must choose standard products from manufacturer catalogues. To do otherwise would incur large surcharges of up to 300% for bespoke production. The design of precast structural and façade elements is carried out by structural engineers and architects working to guidelines given in the precast product literature. The precast manufacturer concentrates solely on the production and supply of units to the site. They use flat bed casting methods that are semi-automated and highly mechanised, employing the minimum of labour to keep costs down. The precast prices are based on high volumes and standardisation of the product range. They are not involved in the site assembly and erection of precast units. That is carried out by the main contractor who is familiar with precast composite construction. The preferred choice of construction of residential buildings in Finland is precast floor planks with precast load-bearing façade panels. When looking at residential architecture in Finland you become aware of the similarity of composition, the standardisation of façade panel construction and how creative architects can be within these tight parameters: a very compelling argument that good design need not be expensive.

By contrast the architectural cladding market in the UK is the total opposite. There is no standardisation of façade elements industrywide or from one project to the next. Architects and designers are free to scheme their layouts, unique to their own project. They are encouraged to use the same panel unit and assembly arrangement to reduce the cost of mould making, but that is often not possible. The façade units are designed by the precast company who usually erect and assemble the units as a total supply and install package. Consequently the precast company will carry a lot more overheads and risk. By encouraging bespoke, non-standard units to be specified, they attract a much higher price in production. Each precast company will have their own connection detail and fixing arrangements. As a result we see exuberance, expressive and flamboyant architecture that comes at a price premium, but there are examples where restraint and rigour has given a fine quality to the structure. They all have one thing in common – they are all different and that perhaps is the telling attraction and appeal of British precast architecture.

As a result of these divergent market conditions, the architecture will differ in scope and aspirations from one country to the next. Precast design reflects the economic constraints on local production as much as the self-conscious attempts by architects to imbue artistic endeavour, context, creative inspiration and ordered formality into the functional purpose of a building. This collection of projects from Sweden, Denmark, Finland, France, Germany, Scotland and England shows how precast concrete in all its different forms, modes and finishes can be brought together creatively and thoughtfully. Some make use of bold vibrant colours and shapes, some draw expression from restraint and tautness of standardised components, while others show how light-weight glass fibre reinforced concrete and the new ultra high strength precast CRC and Ductal® products offer new possibilities in precast architecture.

Each project has been reviewed as a case study with illustrations and descriptions on how it was designed and built and how the precast elements were specified. While the examples are not a definitive list, they have been recognised for their excellence of concrete expression. The section on materials and methods will provide the reader with information on the many different ways to precast concrete and the many choices of surface finish, texture and profiling that are possible.

I am indebted to all the architects and precast manufacturers who gave up their time to share their knowledge with me. I wish to thank those organisations and individuals who helped to make the research to this book possible by arranging my visits to each country. They are BDZ and Jörg Fehlhaber in Berlin, Betongvaru-industrin and Lena Frick in Stockholm, Aalborg White and Hans Bruun Nissan in Denmark, Lafarge Ductal® and Mouloud Behloul in Paris and The Confederation of Finnish Construction Industries RT and Arto Suikka in Helsinki. I also thank Martin Clarke of British Precast for his helpful contacts in Europe and Ian Cox and his team at The Concrete Centre for supporting the book in the UK.

I have learnt so much about precast from researching this book. I hope it brings enlightenment and interest to designers who share an enthusiasm for concrete and perhaps converts one or two sceptics to take a closer look at precast architecture in all it forms. The new ultra high strength materials are sensational.

This book is dedicated to my editor Ria Stein at Birkhäuser who without fuss, formality and bother brings the chaotic and piecemeal arrival of text and images into a coherent, structured and concise work that is then skilfully designed by Alexandra Zöller. Cheers to you both and heartfelt thanks!

David Bennett

PRECAST MATERIALS AND METHODS OF MANUFACTURE

Concrete has been a very versatile and durable material for replicating natural stone for over a century. The increasing scarcity of natural stone and the great expense of cutting and transporting it, has opened up a worldwide market for the production of reconstructed stone and precast concrete using cement as the binder. Fine dust matched to the colour and texture of natural stone is combined in a matrix of fine aggregates, cement and pigments and placed in moulds to form stone-like facing panels, slabs and decorative detailing. In the early years reconstructed or cast stone was processed by the moist-earth or dry cast method where the mix was made semi-dry with low water content and consolidated in timber moulds by ramming or tamping. Modern dry cast stone has a higher porosity than wet cast methods and lower strength, and this tends to limit production to relatively small unit sizes. This method is still used successfully today to replicate both simple and intricate details including ashlar walling, quoins, cornices, sills, string courses and columns on buildings.

The more sophisticated wet cast method of production commonly referred to as precast concrete and the focus of this book, uses very workable, fluid mixes of aggregates, cement and pigments and water. The fluid mix is poured into grout-tight moulds or formwork and compacted by internal and external mechanical vibration and allowed to harden. Precast concrete has high strength, low porosity, low moisture absorption and greater durability. Its fluid consistency allows it to be moulded into complex and intricate shapes. It can be fully reinforced to form large storey-high panels that can be crane-handled, making site installation fast and less labour intensive.

Precast concrete as a structural engineered stone offers new possibilities in expressing the intrinsic qualities of the raw materials – cement, aggregates and pigment. Here the material's plastic form, the choice and range of colours, combined with surface texturing and profiling gives scope and great opportunity to design with freedom and imagination. The surface can be finished with an acid-etch, grit blast, mechanical abrasion or diamond polishing to give it a terrazzo-like appearance.

For integrating precast with high-tech curtain wall systems, the dead weight of the panel can be reduced significantly by specifying light-weight glass fibre reinforced concrete known as GRC. The material is cast in moulds in exactly the same way as precast concrete except that it is reinforced with alkali-resistant glass fibre strands – there is no steel reinforcement – and it can poured in place or spray-applied in layers. GRC panels are easy to handle, they do not require heavy cranage on site and can be installed using a cradle system, they are resilient and do not corrode.

The use of recently developed ultra-high performance concrete in the manufacture of precast concrete offers radically new and dramatically slender structural possibilities in concrete. Two innovations in the ultra-high performance materials have shown how these products can be used to form architectural elements, balcony slabs, staircases and bridge structures that outperform conventional concrete structures and can compete with steel for slenderness.

Dry Cast Concrete

This technique dates back to Roman times where a mixture of lime and pozzolanic cement, sandstone fines and aggregates was made with just enough moisture to hold it together without crumbling. The semi-dry mix was rammed into wooden moulds and left to harden. It was used for making simulated sandstone lintels and for repairing stonework. An example of this can be seen in repair of the Visigoth walls at Carcassonne in south-west France, built in AD 1135.

With the discovery and commercial development of Portland cement in the last century, the dry cast method of producing cast stone was used extensively in the manufacture of artificial stone blocks and facings. It was employed to imitate with great economy, the natural Portland and Bath stone façades of classical Georgian buildings for example and later modelled for Art Deco and Neo-Classical architectural styles. The cast block can be sometimes carved while still green to decorate and sculpture the surface, although such detailing would usually be incorporated in the mould. Cast stone is formed with a semi-dry facing layer comprising a mixture of crushed stone and cement, backed with an ordinary semi-dry concrete layer, which can incorporate reinforcement for strengthening load-bearing elements.

Top: Pediment over doorway (dry cast)
Bottom: Precast quoins corner and decorative features on Shillington Manor, England (dry cast)

The timber mould for forming the cast stone is filled with a 40mm layer of the facing mix which is tamped with an air powered hammer to fully compress the material in the mould. The surface is lightly scratched to ensure an adequate key for the backing concrete which follows in layers of 50mm and is similarly consolidated. Small man-handled pieces which are simple in shape and generally 75mm thick can be de-moulded immediately after the mix has been rammed. The rammed concrete is firm enough to be turned out of the mould without damage. This makes the dry cast method very cost-effective, as one mould can turn out many units per hour. Where delicate ornamental shapes and deep surface profiles are required, a more homogenous cement-rich mix is used and the concrete left to cure in the mould for 24 hours.

Current methods of production are considerably more advanced than the techniques used in the middle of the last century. There are two different production techniques – the first of which is suited to high volume production of small man-handled units. Here the moulds are immediately turned out once they have been filled. The second method which results in far higher quality and detail requires the concrete to be left in the mould overnight before removal. This technique is more commonly used for casting columns, balustrades and architraves and in ornamental landscape artefacts.

Both methods require vapour curing to achieve their optimum strength. The dry cast mix when compressed in the hand will hold together without crumbling and not leave an excessive residue on the hand. The mix of cement and pigment will include very fine crushed natural stone aggregate which has been selected for stone replication, and incorporate a waterproofing admixture such as aluminium stearate, calcium stearate or an acrylic emulsion, to reduce porosity. Most mixes contain coarse aggregates that are generally 3mm in diameter and rarely more than 6mm. Inorganic pigments are used extensively and blended in with either grey or white cement at proportions between 2% to 6% by weight of cement.

Dry cast units will usually require no further surface treatment. Corners and arrises which can be friable should be fully vapour cured. If they get damaged during handling, they should be repaired at the earliest opportunity. The hardened surface can be grit blasted, acid-etched, tooled and traditionally carved. Fixing and detailing of earth-moist units is the same as for natural masonry construction.

Wet Cast Concrete

Increasing mechanisation in construction, the use of the tower crane, the high cost of labour and the need to build quickly created the demand for prefabricated building components and of course large precast façade panels. To form large modern precast panels economically, the concrete mix must have a liquid consistency that will allow it to flow into the moulds without segregating and combine with the reinforcement bars to produce a durable self-supporting structure – hence the term wet cast production

Wet cast concrete, also known as conventional precast, will have a high cement content as it requires a higher water content to create workable, flowing mixes. The proportion and combination of sand, coarse aggregates (up to 20mm in size), cement and pigments will be selected to give the desired finish and will be based on many years experience of precast production. The concrete mix will include the use of water-reducing admixtures and water-proofing agents. It will have been tested for compatibility with the mould oil and formwork face to ensure no adverse effects will arise due to tannins and sugars in the wood, crazing from smooth polished surfaces or staining from release agents.

When the mix is placed in the mould it has to be consolidated using internal and external vibration to remove entrapped air voids and draw the pigment and cement particles to the exposed surface. It is essential that the moulds are made watertight as any leakage of the cement and pigment will leave an unsightly discolouration and honeycombing. In relatively small moulds it may only be necessary to rebate or groove the sides and end of the moulds before they are clamped tight. Larger moulds may need foam gaskets at critical joints or neoprene barriers to prevent grout loss.

Some manufacturers of modern precast concrete prefer to use resin faced plywood or GRP (glass fibre reinforced plastic) lined timber for constructing the moulds. Others will use metal forms because of the high re-use factor for the casting of standardised products. For surface profiling and embossing decorative features, GRP and synthetic rubber liners are placed in the timber moulds. Steel

Dry cast production

Precision and scale of modern precast design (Federal Labour Court, Erfurt, Germany)

moulds can be ideal for casting very large units although they tend to produce a much darker finish with a shiny surface. They cost many times more than the equivalent timber moulds but are capable of being used several hundred times.

The precast moulds are laid flat during concreting; the top face (the reverse side of the panel) is left exposed to be trowelled level after the concrete has been vibrated. This is called flat-bed construction and it is how most precast panels are formed. Sometimes the mould is cast in the vertical position and formwork has to be secured and braced with wailings, props and ties to ensure that it remains rigid and watertight. Such a construction method is specified when the surface has to be heavily grit blasted or point tooled but is more expensive as more formwork materials are required.

The concrete is left in the moulds for at least 18-24 hours to cure and harden before the formwork can be stripped. For economic production, the moulds should have at least 30 uses before being discarded. It is rare to find a building which has 30 or more identical units on the façade. To mitigate the penalty of low repetition, façade panels should be designed as similar shaped units which can be cast from one master mould. Major economising in production cost can be achieved by making small alterations to the master mould.

When the concrete panel is removed from the mould, the facing surface is immediately rubbed down to remove any mould oil stains and surface blemishes and then left to fully harden before further surface treatment is carried out. Occasionally the surface is washed with a cement-pigment paste to fill any air holes and to homogenise the surface appearance. Precast concrete can be finished in a number of different ways which includes acid-etching, applying retarder and water jetting, surface rubbing, sand and grit blasting, bush hammering, point tooling and polishing with diamond or carborundum discs to give a very smooth surface finish.

The maximum panel size that can be precast is governed by two factors: the dimensions of the façade opening and the maximum length and width that can be transported on a lorry. In the UK this is 12m long by 5m wide by 4.1m high for a lorry travelling on a motorway with police escort. If the width is restricted to 2.89m such a load can travel on any road in the UK without police notification.

Thermal insulation can either be fixed to the back of the unit on site after the panel is installed or factory applied. Composite precast sandwich panels with thermal insulation in the core are manufactured in some European countries, where the problems of cold bridging have been overcome by the use of proprietary anchors.

Large precast panels are usually formed with an integral nib which sits on the supporting floor slab or beam, with the top of the panel pinned to the main structure to allow for differential movement. Joints between panels are sealed using silicone or polysulphide sealants.

Mould for curved panel being bolted together

Forming timber master mould

Light acid washing

Self-Compacting Concrete (SCC)

Where difficult vertical sections of precast are required with very congested reinforcement, then a self-compacting concrete is often specified. For example the massive charcoal grey free-standing precast blocks for the Jewish Memorial in Berlin designed by Eisenman Architects – up to 3m high and weighing up to 20 tonnes – have been cast with pigmented self-compacting concrete.

Self-compacting high performance concrete minimises air voids induced during the placing process and those formed because of the excess water required for workability with normal compacted concrete.

By introducing a viscosity agent in the mix, the viscosity of the concrete paste can be increased effectively to inhibit segregation. For practical necessity, the proportion of fine to coarse aggregates is kept at 1:1 by volume. If the coarse aggregates exceed a certain limit, there is greater contact between the larger particles which increases interlocking and the risk of blockages on passing through spaces between reinforcing bars. The possibility of interlocking is negligible if the solid content of the coarse aggregate fraction is lower than 50% of the total mix, provided adequate mortar is used. Smooth, rounded river gravels are generally preferred because this permits a larger coarse aggregate volume than an angular or rough textured one.

Fine aggregate is defined as particles that are larger than 90 microns, anything smaller is defined as powder. The amount of water confined by the fine aggregate is almost proportional to the volume of fine aggregate, so long as the fine aggregate proportion is around 20%. Selection of the powder is critical because the properties effect self-compaction and govern the quality of the hardened concrete. SCC mixes contain higher than normal proportions of fine material that are smaller than 90 microns such as pfa (pulverised fuel ash), ggbs (ground granulated blast furnace slag) or limestone powder. One of the characteristic features of a powder is that a unit volume confines a large amount of water. There is an optimum water/powder ratio for imparting viscosity to the mortar paste for self-compacting concrete.

Properties of hardened SCC do not differ significantly from those of ordinary concrete of a similar basic composition. For compressive strength compliance the standard concrete cube test will be adequate. The extra cost of the higher cement content and special admixtures can be justified by the greater savings on the labour and time consuming activity of vibrating concrete. The elimination of compaction opens up the opportunity for greater concrete automation into the production process.

Sandwich Panel Construction

One of the most structurally efficient product options is sandwich panel construction. The panels are storey-high units up to 8m long with an outer façade panel of between 60mm and 80mm thick that can be finished with a wide range of surfaces; a layer of insulation and then a backing leaf of either load-bearing or self-bearing precast concrete between 90mm and 120mm thick. Self-bearing means that the sandwich panels can only support their own weight. When the inner leaf is load-bearing it can support the structural floor and the façade above it.

The structural floors are usually precast hollow core planks that are stitched to the top of the inner panel. The load-bearing sandwich panel offers many advantages. They are fast to erect, they eliminate the need for columns and wet trades, they are self-finished and the most competitively priced residential building system in Europe.

The two layers of the panel are interconnected by steel ladder reinforcement which act as wind and shear connectors. The thermal bridge through the steel connectors is minimal. The system has the advantage of providing structural integrity without placing any reliance on the insulation for load transference.

Sandwich panel details: outer leaf decorative, insulation layer and load-bearing inner leaf

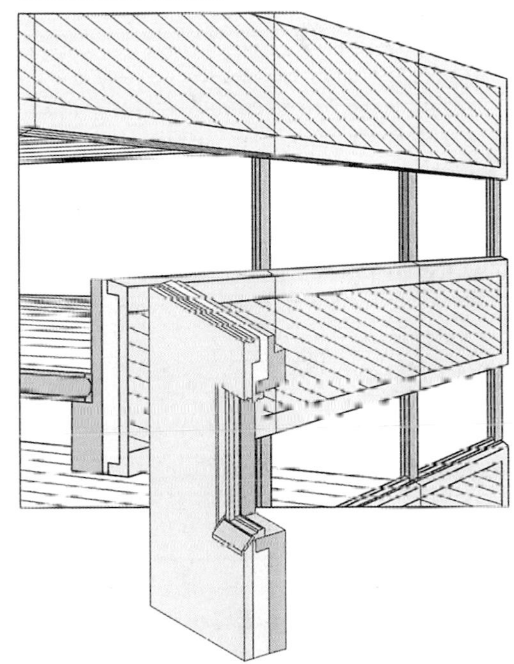

Typical sandwich panel construction

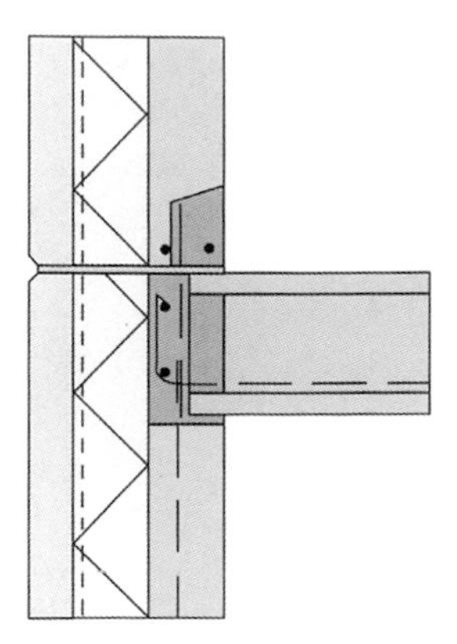

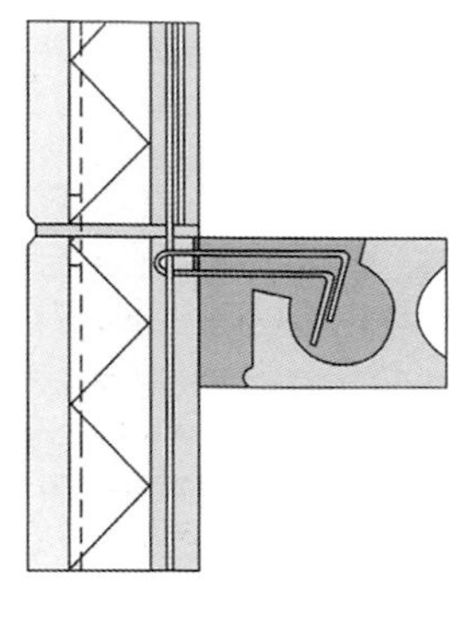

Typical sandwich panel design

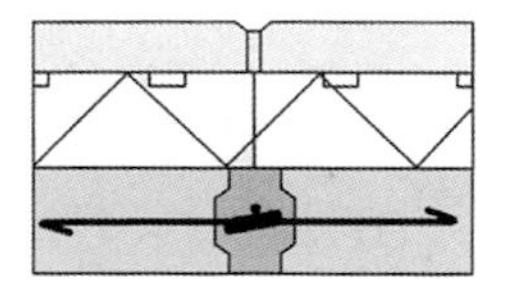

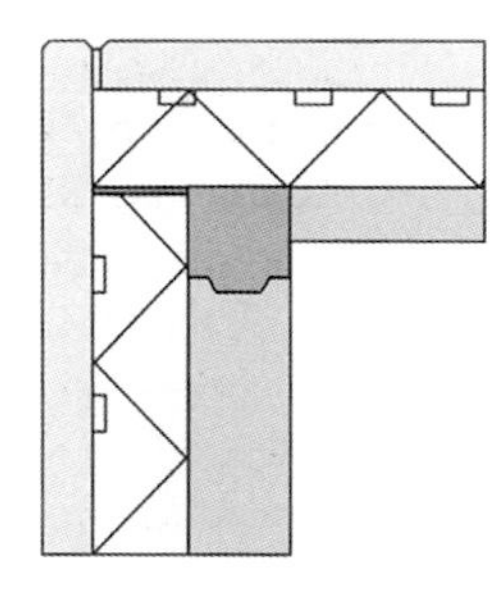

Connection details, left to right:
Floor to façade
Side wall to façade
Corner solution

Precast Light Concrete

The trend in modern building construction towards more lightweight 'high-tech' façades using glass curtain walling, resin coated aluminium and steel fascia, has disadvantaged the heavier precast and reconstructed stone cladding unit. In the 1970s a lightweight precast cladding system developed by Pilkington and the BRE (Building Research Establishment) in the UK offered the same range of surface finishes as conventional reconstructed stone and precast but with a considerable reduction in panel thickness, from 150mm down to 20mm. Saving in panel weight was due to the strength and toughness of alkali resistant glass fibre strands that reinforced the cement mortar mix. Typically the skin thicknesses of the panels are between 15mm and 20mm, making its weight as much as 80% lighter than the corresponding precast concrete unit. Weight reduction of this magnitude offers substantial savings in transportation, handling and site erection cost. Glassfibre Reinforced Concrete or GRC is composed of a mortar mix of cement, selected crushed aggregates, sand, fillers, admixtures and water and alkali resistant glass fibre strands. The glass fibre is typically between 6-51mm long and 10 to 30 microns in diameter. It obtains its alkali resistance from a coating applied over the glass strands in the manufacturing process. For sprayed GRC it is recommended that 5% fibre by weight of total cement mortar should be contained in the mix, to achieve optimum tensile strength. Combinations of fibre lengths varying from 6-51mm and depending on the production process, will ensure that adequate bond strength develops between fibres and the cement matrix and encourage a quasi-ductile failure by fracture of the fibres.

Examples of GRC panel designs

Hand sprayed GRC panels are usually produced with a water/cement ratio lower than 0.4 (low in comparison to most concrete) and a cement content of not less than 800kg/m^3 (high for concrete). A typical hand sprayed GRC mix produced by Trent Concrete contains around 5% of glass fibre strands. A fully automated spray system for manufacturing GRC façade panels operated by Durapact in Germany uses fibre lengths of between 12-50mm at the dosage of 5% by volume of the concrete mix.

GRC is a very mouldable and adaptable material for forming architectural features such as cornices, parapets, quoins and other imitation stone features to dress building façades. As a lightweight cladding system for whole buildings it has gained a limited foothold in the façade market, due to some minor teething problems associated with brittleness, differential movement and crazing when it was first launched.

The development of more durable glass fibre strands, extensive testing of the product over many years and improvement to the mortar mix to reduce long-term brittleness, has given specifiers the assurances and compliance standards they require. The various techniques used to manufacture GRC products – manual and mechanised spray methods and traditional wet casting – allows the material to be formed in a wide variety of shapes and profiles. It can be moulded easily to suit classical or modern architectural expression using complex profiles and curved or angular surfaces. Being cement based with no metal reinforcement, it has inherently good durability and chemical resistance. It is non-combustible and has high impact strength. It is not susceptible to rust staining or corrosion, and can be used in combination with insulating material and sound proofing. Constraining factors in performance are generally due to its relatively large thermal and moisture movement and low ductility.

In designing GRC two characteristics are important, the LOP value 'The Limit of Proportionality' and the MOR value 'Modulus of Rupture' which are derived from accelerated ageing and bending tests. Typical material properties for GRC panels can be determined from test data published by the GRCA in UK and similar organizations in Europe and shown in their guidance notes. This includes compressive strength, modulus, impact strength, poissons ratio, LOP and MOR values, ultimate tensile strength and shear strength. With such a high cement content and low water/cement ratios GRC panels are more resistant to surface weathering, moisture ingress and discoloration than reinforced concrete cladding panels.

The need for GRC cladding to be flexibly mounted on the supporting structure to accommodate thermal and moisture-induced movement is therefore important. Many of the problems associated with GRC have resulted from the rigidity of the fixing, from errors of installation or from introducing some other unintended restraint to panel movement. Wherever possible, design GRC panels as independent skins to allow maximum freedom to shape, curve and profile panels. Good detailing of panel size, reducing horizontal flat surface areas like window sills, which may collect surface water and create high moisture gradients in the panel, and avoiding panel shapes that wrap around a building corner causing large thermal movements, will ensure a longer service life.

A wide choice of mould material is available to achieve different surface textures, although timber and glass reinforced plastic are more commonly used. Colour and texture can be obtained by the use of fine aggregate facing material. Crushed rock, sands and gravels are usually employed in combination with white cement and pigments to produce reconstructed stone finishes. The finished face can have the aggregates exposed by acid washing, grit blasting or using retarding chemicals. Such finishes are generally similar to those obtainable with conventional precast concrete; however, with GRC, the maximum aggregate size is restricted to 10mm.

Top: Rolling and compacting GRC
Bottom: Spraying GRC

Precast Ultra Concrete

The growing emphasis on non-combustibility of raw materials, the rise in cost of hydrocarbons and high energy cost of production, led researchers back in the 1970s to look for the possibility of making a 'defect free' cement as a replacement material for organic plastics and aluminium. The major attraction of cement was simply the huge energy savings in manufacture. To produce 1m^3 of cement, organic polymer or aluminium requires 10, 100 and 1000 GJoules respectively of energy. Clearly cement has a major advantage in energy saving because it hydrates with just the addition of water under normal air temperatures. There is no need to heat it. The big disadvantage is its low tensile and bending strength, and low fracture toughness due to the air-voids trapped inside. Removal of the internal voids, air pockets and capillary pores – the defects – from the 'cement paste' was made possible by introducing a small proportion of water soluble polymers which reduced inter-particle friction and surface tension and made the cement particles pack much closer. The increase in compressive strength of up to 300Mpa was phenomenal but its application was limited to injection moulding and extruding and not concrete structures.

Researchers in Denmark and France in the last ten years have developed ultra high performance concretes which can be precast in factories and poured in place just like conventional concrete. Ultra high strength from 100Mpa-350Mpa is achieved by controlling the particle-size distribution of the cement to minimise the void spaces between the cement grains. Some products use just neat cements, polymers and fine fillers with a combination or combinations of plasticisers and superplasticisers to make these earth dry mixes workable, as the water/cement ratio is extremely low at 0.2-0,3. Other ultra high performance products add very finely divided silica sand, filling the voids left by the cement particles. There is no coarse aggregate. The silica has the added advantage of reacting chemically with the cement paste to become an integral part of the matrix. A dispersion surfactant or superplasticiser is also necessary to achieve workability for placing as it is an earth dry mix with a low water/cement ratio.

Ultra high performance concretes are used increasingly for a range of structural applications, and standards in a number of countries are being revised to accommodate these improved materials. Often they are more brittle than conventional precast concretes, which can lead to problems in failure mode as well as under service conditions. One way of overcoming this problem is to provide ductility by incorporating steel fibres in the matrix.

The material properties are far superior to ordinary concrete in every respect. Apart from their high compressive strength, the bond and shear strength is far greater, Young's Modulus is much improved, while the tensile strength and ductility are enhanced with the introduction of steel fibres. Precast ultra materials are structural grade material for precasting slim lightweight load-bearing façade elements, thin cantilever balcony slabs, elegant staircases and supporting columns and remarkably slender bridge structures.

The two products highlighted are CRC (Compact Reinforced Composite) which is a fibre reinforced cement with strengths ranging from 150 to 300 Mpa, developed by Aalborg Portland and marketed by CRCTechnology in Denmark. The other is Ductal® developed by Lafarge and Bouygues in France. They both belong to a special group of ultra high performance fibre reinforced concretes.

CRC staircase

What is CRC?

One formulation of CRC (Compact Reinforced Composite) produces very high bond strength just like superglue, due to the large content of micro silica and steel fibres that can be added to the concrete mortar mix. It can glue together joints between two reinforced concrete sections, in many ways similar to welding steel sections and has been called 'weld cement'. When it is specifically required for jointing structure the formulation is called JointCast. Reinforcing bars need only have a bond length of eight times the bar diameter for a full tension lap as opposed to 30 or 40 diameters for a conventional concrete. For practical reasons the minimum gap between joints are usually 80mm or 100mm wide to be able to pour and compact the JointCast mortar.

In precast ultra applications, CRC is modified to produce strengths between 100-300Mpa according to project requirements. It is supplied to precast manufacturers as a dry powder with the special CRC binder. CRCTechnology will help the producer find a suitable local sand and to source steel fibre to add to the

CRC circular balustrade

binder. Water is added to the mix under controlled conditions. The high strength of CRC with its mortar-like consistency allows for very close rebar spacing, making it possible to precast thin, lightweight structural elements such as balcony slabs and staircases. The steel fibres are necessary to maintain the ductility of the material but are not sufficiently robust to support applied loads nor control deflection, therefore reinforcement is necessary. Typical proportions for making 1m^3 of a 200Mpa mix run as follows:

- CRC binder 1,000kg
- sand 1,290kg
- steel fibre 180kg
- water 150kg

A conventional pan mixer can be used for mixing CRC. The mixing time is between 5 and 8 minutes after the water is added. The bundles of steel fibres are added part way through the mixing. The material is sensitive to temperature change as the high superplasticiser dosage tends to retard the hardening, and should not be used below 5° C. At 20° C the initial set will start after 7 8 hours, with the compressive strength of 60Mpa achieved after 10 hours. Because of its very low water content, in hot weather and drying winds the surface should be covered as quickly as possible to prevent evaporation, and the structure enclosed by tenting with tarpaulins. Surface finishing is a problem as the mix is very sticky. A spike roller used for screeding is quite effective for levelling the surface.

Where CRC has been engineered to its full potential and a balcony slab made to cantilever the span without need of a supporting column, contractors have found it to be the cheapest option. In Denmark there are now three precast companies all doing quite well offering CRC precast slabs, staircases, manhole covers, slender beams, load-bearing columns and façade panels. While compressive strength and ductility is greatly enhanced, stiffness is only slightly higher than for normal concrete, which means that deflection has to be carefully considered and controlled by reinforcement. Allowances have also to be made for drying shrinkage due the high cement content and that is catered for by detailing additional reinforcement.

The CRC binder is more expensive than cement and the steel fibre content also leads to a higher price for CRC. On the other hand, elements are typically one third the volume of conventional concrete. This means that for the applications used in Denmark – typically balcony slabs and staircases – the price for CRC is equivalent to alternatives in steel or concrete. The price of CRC elements can even be lower than alternatives in steel or concrete, when a cantilevered CRC slab replaces a conventional concrete or steel balcony slab supported on columns.

Several Danish architects have produced spectacular design such as the Spiral Staircase in Copenhagen documented as a case study of CRC applications.

CRC staircase flight

Balcony panels

CRC balconies

CRC balconies

What is Ductal®?

Ductal® is a concrete composed of cement particles, fibres, special fillers and plasticisers that is able to fully hydrate with the minimum of added water. It has a water/cement ratio of just 0.2. It is the outcome of over ten years of collaborative research between Lafarge the material manufacturer, Bouygues the contractor and Rhodia, a chemical manufacturer. Through intensive research and development work, the material has been patented, refined and commercialised. Fifteen universities and six testing laboratories in different countries have also contributed to the research effort.

In May 2002 design guidance rules and material recommendations were formulated in France for the use of Ductal® in structural applications. These recommendations were established by a working group comprising representatives from leading construction companies, building control agencies, suppliers, certification authorities and coordinated by SETRA (Road and Traffic Government Agency). The material consists of cement and cementitious fillers carefully blended and graded, with a particle size distribution ranging from a maximum of 600μm (0.6mm), down to less than 0.1μm to obtain the densest packing with the minimum of void spaces. It has sand fines but no coarse aggregates. It is a super high strength concrete mortar with the minimum of internal and external imperfections such as micro-cracks, air voids and pore spaces. This enables the material to achieve a greater percentage of its ultimate load-carrying capacity and enhances its durability properties.

The material has a compressive strength ranging from 200Mpa-350Mpa, but lacks sufficient ductility. The inclusion of steel fibres drastically improves the tensile strength and provides a substantial level of ductility. The various formulations and applications of Ductal® are based on an optimisation of the material composition with steel and polypropylene fibres. For example, to enhance its structural performance steel fibres are included and the material is also heat treated to reduce creep and shrinkage strain. For every application the technology can be adjusted to achieve the optimum performance required.

For structural grades, Ductal®-FM is prescribed with steel fibres, for a smooth decorative material that can be handled Ductal®-FO is recommended which has polypropylene fibres, while for enhanced fire resistance Ductal®-AF is prescribed which has a combination of steel and polypropylene fibres.

The fresh mixing of these materials as well as the controls that have been developed, makes it relatively easy to handle in terms of flow and self-compaction. With minor adjustments, most conventional concrete batching equipment is suitable for mixing it. The matrix gives a very fine 'bone china' surface finish that can be moulded to replicate any kind of profile or intricate pattern. By using adequate pigments a range of coloured concretes can be achieved for architectural and decorative applications.

A typical load-deflection curve for Ductal® under a three point loading test is shown in the drawing. The material exhibits linear elastic behaviour up to first crack and has considerable ductility thereafter until the ultimate flexure load is reached, where upon it begins to yield with plastic failure until rupture. It has an ultimate bending strength which is over twice its first crack stress and more than ten times the ultimate stress of conventional mortars. With such high strength and ductility structures can be designed without any secondary passive reinforcement and no shear reinforcement.

The use of this concrete-like material has almost unlimited possibilities of appearance, texture and colour. It has excited architects by giving them access to an unexpected new world of shapes and forms. Ductal® has been used in architectural applications like the bus shelters in Tucson (USA), façade panels in Monaco and the Kyoto clock tower in Japan. A number of footbridges have been constructed using Ductal®, they are the Sakata Mirai in Japan, the bridge in Sermaises in France, the one in Sherbrooke in Canada and the Seonyu Footbridge in Seoul, highlighted as a case study.

Examples of Ductal® applications:
Brise soleils on apartment building
Curved roof for train station in Calgary
Sakata Mirai footbridge, Japan

Ductal® load-deflection curve

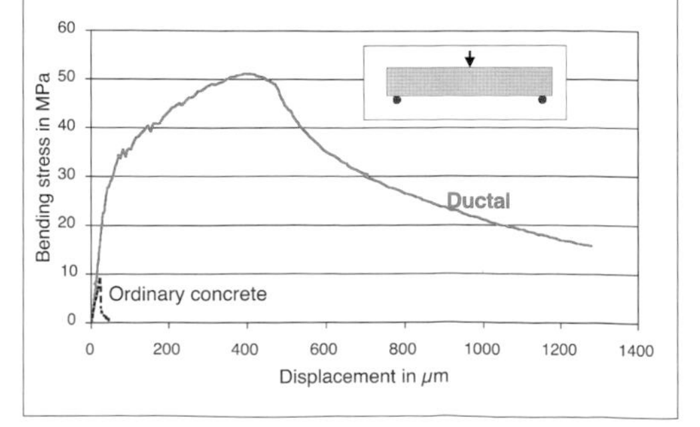

Types of Surface Finishes

Precast concrete like no other modern material can be moulded into the most extraordinary shapes and forms and for each of these shapes the surface can be expressed by colour, texture and profiling. The possibilities are endless, the choices are myriad, limited only by the practicality of casting it and the cost.

Plain Cast Finishes

Natural: For a plain concrete finish the mechanism that gives colour to concrete is the light absorption qualities of the finest particles that migrate and saturate the surface. For ordinary concrete mixes, the cement particles are the finest particles and therefore the colour of the cement will dominate the surface colour. The final colour of plain concrete depends on the cement colour, the water/cement ratio – the higher the ratio the lighter the tone, the lower the ratio the darker the tone for the same cement – and to a lesser extent the very finest particles in fine aggregate that are less than 50 microns.

Pigmented Concrete: If a pigment was introduced into the mix, the pigment colour will dominate because it is much finer than cement. The amount of pigment needed to colour a concrete will vary according to the cement content, the pigment type and the method of incorporating it into the mix, but it is usually between 3-6% by weight of cement. Pigments with white cement create bright colours, and pigments with grey cements create duller earth colours.

Naturally occurring pigments are inert oxides and hydroxides of iron and titanium and copper complexes of phthalocyanine, found in mineral rocks. They range in colour from red, brown to yellow. The full description of pigments specified for use in concrete or mortar are given in current standards. A pigment may be defined as a fine dry powder or an aqueous suspension or slurry of powder, virtually inert to the ingredients of the concrete. The mineral rocks containing raw pigment deposits are quarried, heat treated, crushed and then ground to a flour consistency to create industrial pigments. And like sands and coarse aggregates pigments have unique characteristics – some are needle-shaped, some are spherical, some are much smaller than others, while some like the phthalocyanines are hydrophobic. The various pigment mixes of red, yellow and brown oxides are blended to create intermediate colours, and have to be carefully batched so that the bulk density and water absorption are known and can be adjusted in the mix.

It is for this reason that synthetic oxide pigments were developed by Bayer to create a more homogeneous pigment particle size with a more uniform bulk density and water absorption. Synthetic pigments are more intense in colour than organic pigments, and have excellent long term colour stability. The pigments are produced in basic colours – red, black, and yellow – by Aniline or Penniman-Zoph process. In the Aniline process nitrobenzene is reduced to aniline in acid solution using fine iron filings as the reducing agent. During this process the iron filings are oxidised to produce an iron oxide which eventually turns a blackish grey in colour. By controlling the oxidation it is possible to produce black and yellow slurries with a high tinting strength. After washing and filtration the slurry is dried out to produce black and yellow pigments or heat treated and calcined to produce red oxide pigments. In the Penniman-Zoph process, iron filings captured from scrap sheet metal are dissolved in acid solution in a hydrolysis process, involving the oxidation and hydrolysis of iron sulphate in the presence of metallic iron; this produces an iron oxide yellow pigment with needle-shaped particles. A range of brown pigments is blended from these three basic colours. Green and blue pigments are processed from copper oxides and cobalt deposits and are very expensive. Synthetic pigments are generally preferred for all architectural concrete work.

Cleaning edges of precast panel before despatch

Using synthetic oxide pigments in concrete with a high tinting strength means that colour saturation is usually achieved at 5% pigment/cement ratio with white cement. Any higher dosage will not increase the intensity of colour of concrete. For most coloured concrete production, pigments are introduced into the mix by dispensing them with the mixing water or through a plasticising admixture suspension. One of the big headaches of coloured concrete and mortar production has been the problem of lime bloom or efflorescence caused by the carbonation of calcium hydroxide that migrate to the surface where it forms white deposits. Besides a good concrete mix design with a low water/cement ratio, prevention of rapid drying out of concrete in the early days was the best way to eliminate secondary efflorescence which can occur throughout the life of the concrete, until it has fully carbonated. A surface coating of a transparent vapour permeable membrane that is none-yellowing and does not break down under UV light would be helpful.

Colour Fastness: Experienced precast companies in Northern Europe all agree that it is impossible to maintain the original shade of a pigmented concrete – especially dark grey or black concrete – as the colour fades with time and can be different when cast in winter or summer due to the prevailing air temperatures and humidity. To ensure the most consistent colour, white cement is always specified as the cement colour does not vary. If an ordinary grey Portland cement was used or blast furnace slag cement specified there is a serious risk that the colour of the cement may change over the year due to subtle variations in the raw material which affect the final colour.

As concrete hydrates it also carbonates since CO_2 combines with the free lime in the cement to form calcium carbonate. This has the effect of lightening and slightly bleaching the concrete surface colour. The effect is more evident on dark concrete backgrounds. There is also the risk of white deposits of efflorescence forming on the surface which will fade with time. To minimise this risk precasters acid-etch the surface. To ensure a rich and stable concrete finish it is often best to expose the naturally coloured coarse aggregates and incorporate crushed aggregate fines in the mix to naturally colour the mortar surrounding the coarse aggregates.

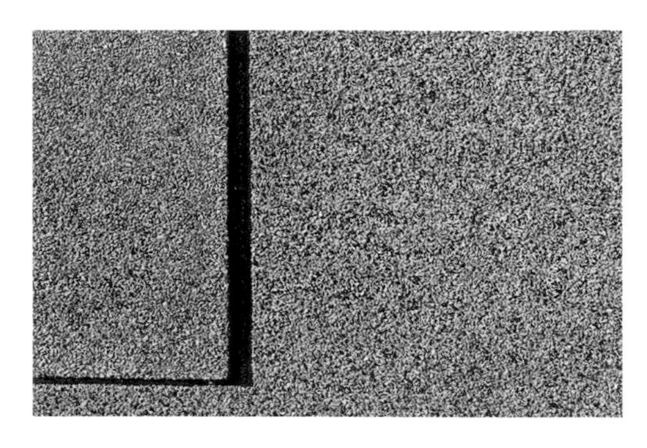

Top upper: Exposed black basalt aggregate finish
Top lower: Exposed aggregate façade (VTG Building, Berlin)

Exposed Aggregate Finishes

Surface Retarders: This is applied when the concrete is fresh and it prevents or retards the surface skin of cement from setting. When the concrete has hardened the surface is water jetted to expose the aggregates to a depth of 2-5mm depending on the strength of retarding agent applied to the contact formwork. The surface is cleaned with dilute hydrochloric acid to remove traces of lime that can smear the aggregate surface.

Aggregate Transfer: A variation of surface retarding is that of aggregate transfer especially where large pieces of aggregates are required. The aggregates are placed face down on a thin bed of sand in the mould. At least half the depth of the aggregates remain uncovered for the concrete to harden around them and retain them firmly in the panel. When the concrete sets and the mould is removed the aggregates are washed to remove the sand. Some precasters prefer to hold the aggregates on a liner which is then placed in the mould. When the concrete hardens the soft liner is removed on the exposed face and the aggregates are washed.

Acid Washing: Hardened concrete can be acid washed with dilute hydrochloric acid to remove the surface laitance without exposing the coarse aggregates. The surface has to be thoroughly washed afterwards to remove all traces of acid residues to avoid subsequent staining. Operatives will need to wear protective clothing against accidental spillages. For small panels the acid is brushed over the surface to etch it and then washed over with water. Larger panels are carried by gantry crane to acid tanks and immersed for a set time then washed down by water spray. Acid will attack some aggregates, for example limestone and marbles, altering their surface texture which may impair their surface quality if it is a deep etch.

Bottom upper left and right: Acid washing
Bottom lower: Grit blasting

Grit Blasting: Depending on the pressure and grit size used, it is possible to achieve a variety of different surface finishes, from a light abrasion which removes just the surface laitance down to a deep etch to reveal the coarse aggregates. The grit used in blasting is made from processes mineral slag or metal fibres. The use of natural silica sand is prohibited for health reasons as it is harmful if inhaled, unless the operative is fully protected and the work area is sealed off. The abrasive grit for concrete is chosen according to the particle size – fine for removing the surface laitance, medium for light blasting and coarse for heavy blasting. Water is often introduced into the air jets as a means of reducing the dust created.

Hand polishing

Concrete Terrazzo: The aggregates are exposed to disc abrasion and polishing to achieve a smooth hard wearing surface. Successive use of coarse and medium carborundum and diamond studded abrasion discs cut into the hardened concrete surface to a depth of about 2-4mm to reveal the aggregate matrix. The surface is then polished until smooth. It is more efficient when water is used with the cutting discs; this is known as wet grinding. For large precast units semi-automatic rack mounted machines are used in the cutting and polishing process. Handwork is for small units and for finishing the corners and edges of large panels.

Bush hammering and point tooling power tools

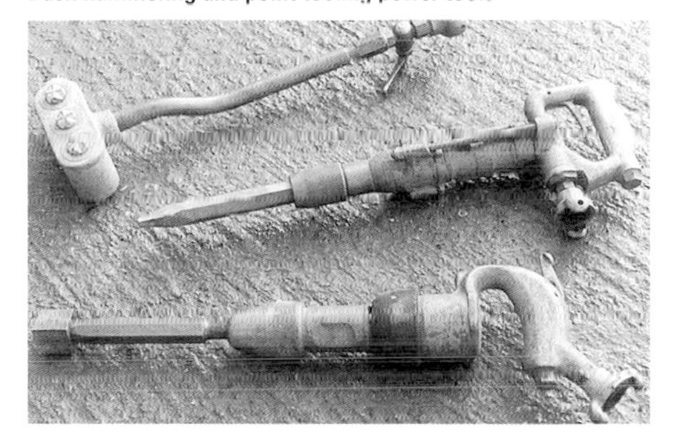

Bush Hammering: The surface is impacted with percussive hammering using electrically or pneumatically driven power tools. The bush hammer can have a 50mm diameter head faced with small raised hard steel cones or a triple head hammer with cruciform shaped heads. The tool has to be placed square to the work face and moved evenly across the surface to impact the surface to the same depth. A layer averaging from 1-2mm is removed fracturing the coarse aggregates and often enhancing their appearance. Point tooling and needle gun are variations of bush hammering abrasion.

Surface Profiling

The ability of concrete to assume the character of the formed face and to be shaped and moulded can be exploited by profiling, using natural timber patterning and synthetic material contouring.

Timber Patterning: This includes rough sawn boarding, smooth planed boarding, plywood grain patterning, cork mats, chip board and wood wool imprinting. The main advantage of this type of finish is that the material is relatively cheap but it has a limited number of re-uses due to its absorbency. It is liable to soften and delaminate with successive casting therefore it is important to waterproof the timber with a proprietary clear lacquer or varnish to maximize the number of re-uses. A carefully chosen release agent will ensure that the finished surface is blemish free with no resin or tannin marks from the timber.

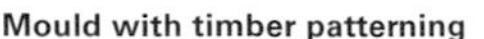

Form Liners: The materials in this category include rigid polystyrene and glass fibre liners and flexible elastomeric liners. For a limited amount of repetition and one off use, the rigid liners are best as they are inexpensive. They are usually made of polystyrene for one-off use and hard PVC plastic and glass fibre for a modest number of repeated usage. Elastomeric liners are flexible and easily mouldable, giving very good definition to the finished concrete with repeated usage of up to 100 times. The liners are fitted over the formwork in the moulds. They are made from plasticised PVC and polyurethane and are expensive. Release agents suitable for polyurethane liners have to be wax based emulsions, as other types can damage the material.

Mould with timber patterning

Examples of precast surface finishes
Left: Plain white panel design (BDZ Building, Berlin)
Upper centre: Surface profiled panel with partial bush hammering
Lower centre: Contrasting plain and acid etch finish
Right: Highly profiled pigmented panel design (Broadwalk House, London)

Support and Fixings

The primary purpose of the fixings is to support the weight of the precast panel and to restrain it at the corners from out of plane movement caused by wind force. The size and weight of the panel and the construction of the building will determine the design and location of the fixings.

The support of the panel is provided by load-bearing fixings which transfer the weight to the structural frame. The support can be a concrete nib cast integrally with the panel that sits on the structural floor or steel beam. It is better to support the panel at or near their bases rather than top-hang them, so that they remain in compression. To reduce the tendency for the panel to tilt outwards the support area should be in line with its centre of gravity. Restraint fixings are to resist wind loads and allow adjustment of the panel for alignment in all three planes. Restraint fixings are located at the corners of the panels and allow adjustment of typically +/-25mm in each direction.

There are many fixing arrangements to support and restrain precast panels and each precast manufacturer will have his own system. There is greater standardisation of fixing arrangement and construction detailing across Scandinavian countries. This is not the case in the UK where precast design is considered a bespoke process, and every building façade treated as a unique event and a one-off activity. Early discussion with the supplier is critical during the design development to ensure compatibility of the support system with the structural frame.

Lightweight GRC panels require freedom of movement to avoid restraint loading on the panel. There is merit in using fixing devices which are not dependent on precise positional assembly by installers and have mobility for adjustment once in position.

Flexible fixings and support details for GRC panels

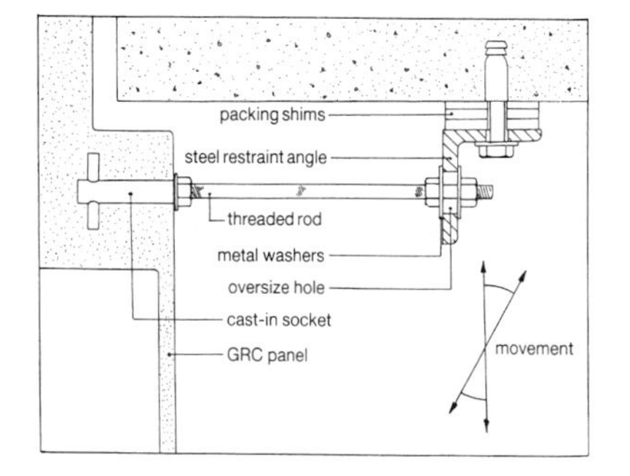

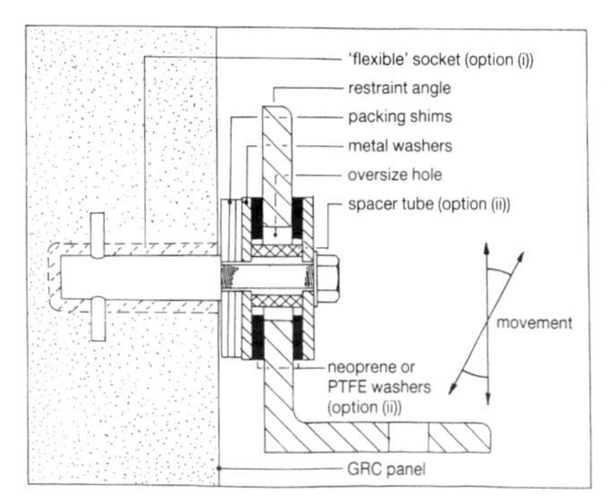

Typical GRC panel with upper load-bearing fixings and lower location fixings

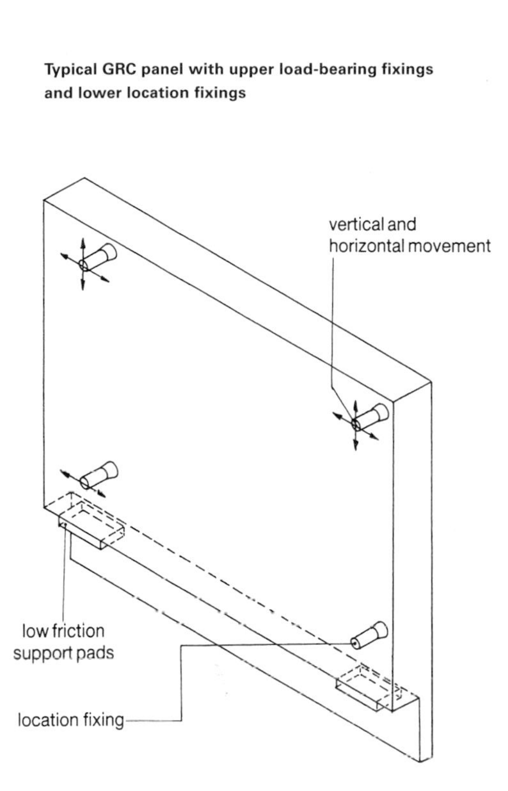

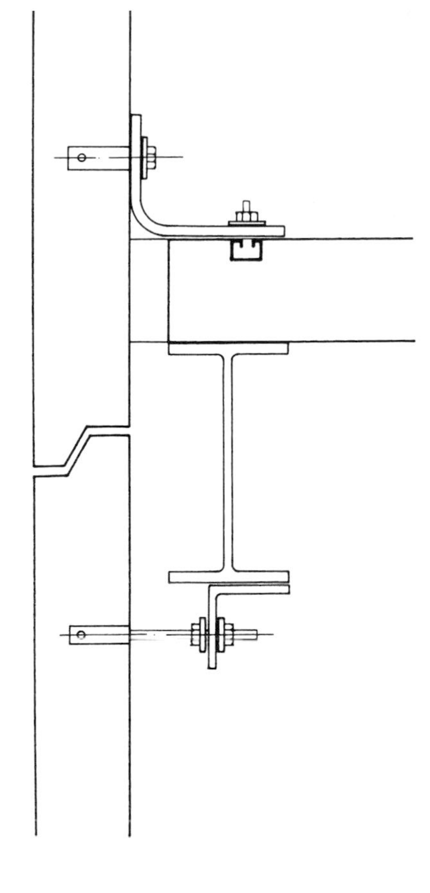

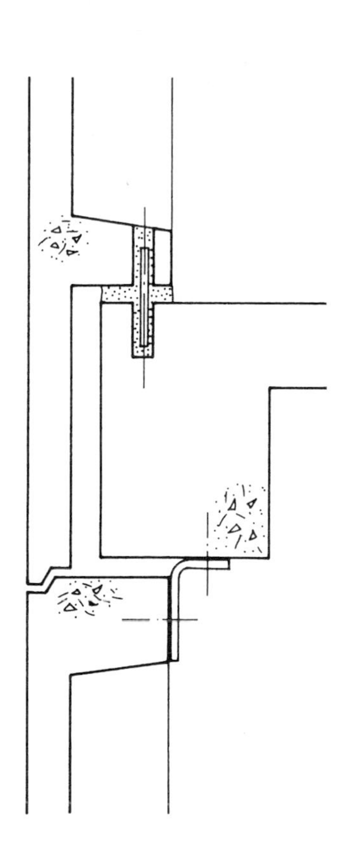

Conventional precast panel support details

Cost and Construction Matters

Precast concrete offers the opportunity to prefabricate the external façade, the floors and the frame under factory conditions. The size of the units, the location of window openings, the joint details, water run-off and weathering, the frame, the support arrangements, cranage capacity, site access and storage are important factors which influence on the detailing of the precast unit. The most economic design results from using panels as large as possible which have a high degree of repetition with at least 30 identical casts from the same mould. Practically this can be achieved by specifying standard components taken from a manufacturer's precast catalogue or involving the precast supplier in the design development stages to make it so. This is the only way to bring about a creative design in precast concrete at a price that is competitive and a construction period that is speedy.

Architectural precast concrete has limitless possibilities if designers work with the suppliers in a spirit of cooperation and understanding and mutual support. The results that are sure to emerge will be on a par with the fine buildings highlighted in this review.

Literature

David Bennett, *Innovations in Concrete*, London: Thomas Telford, 2002.

Susan Dawson, *Cast in Concrete*, Leicester: The Architectural Cladding Association, 2003.

Friedbert Kind-Barkauskas, Bruno Kauhsen, Stefan Polónyi, Jörg Brandt, *Concrete Construction Manual*, Basel: Birkhäuser, 2002.

Christoph Mäckler (ed.), *Material Stone: Construction and Technology for Contemporary Architecture*, Basel: Birkhäuser, 2004.

H.P.J. Taylor (ed.), *Precast Concrete Cladding*, London: Edward Arnold, 1992.

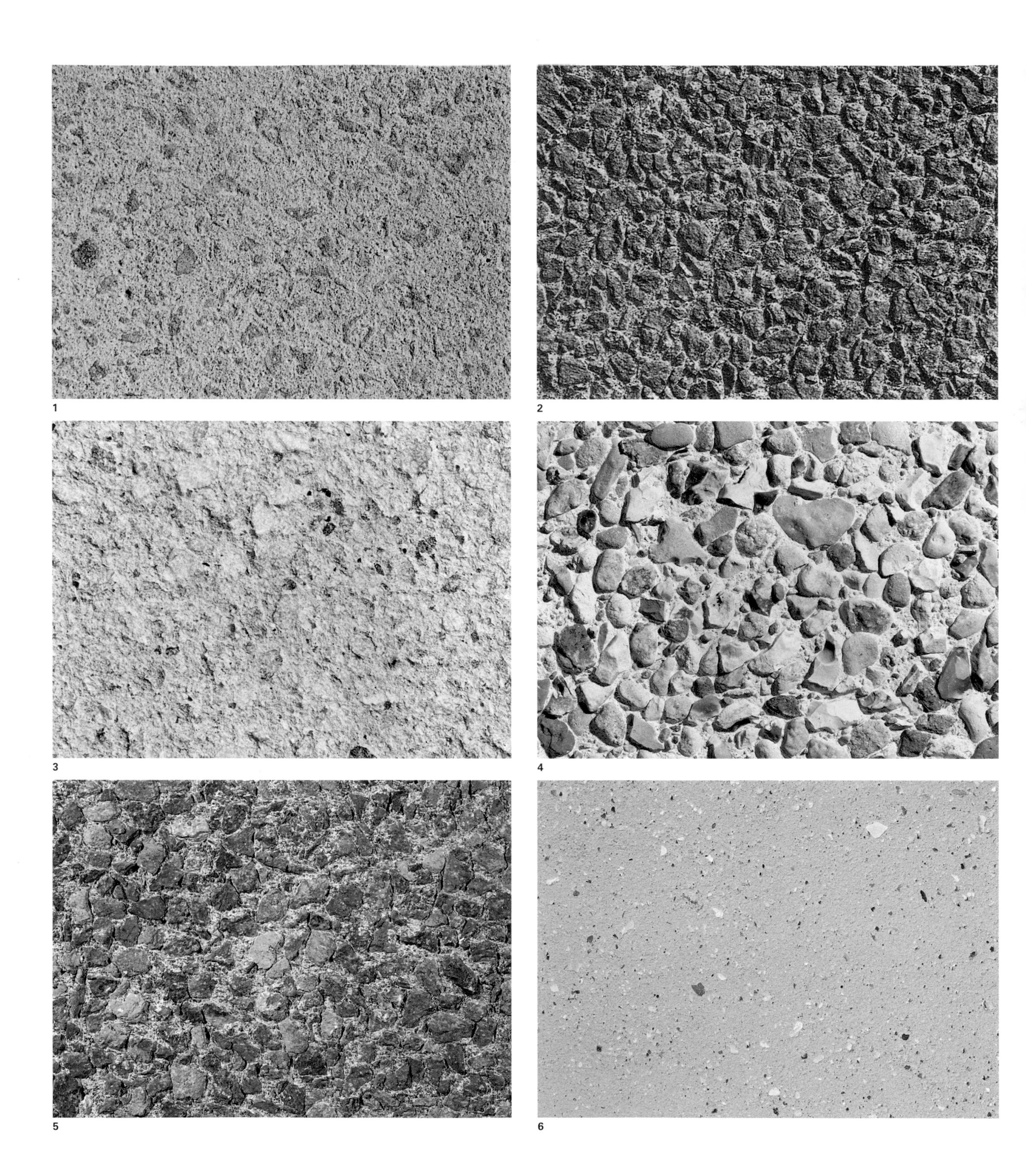

1 2 3 4 5 6

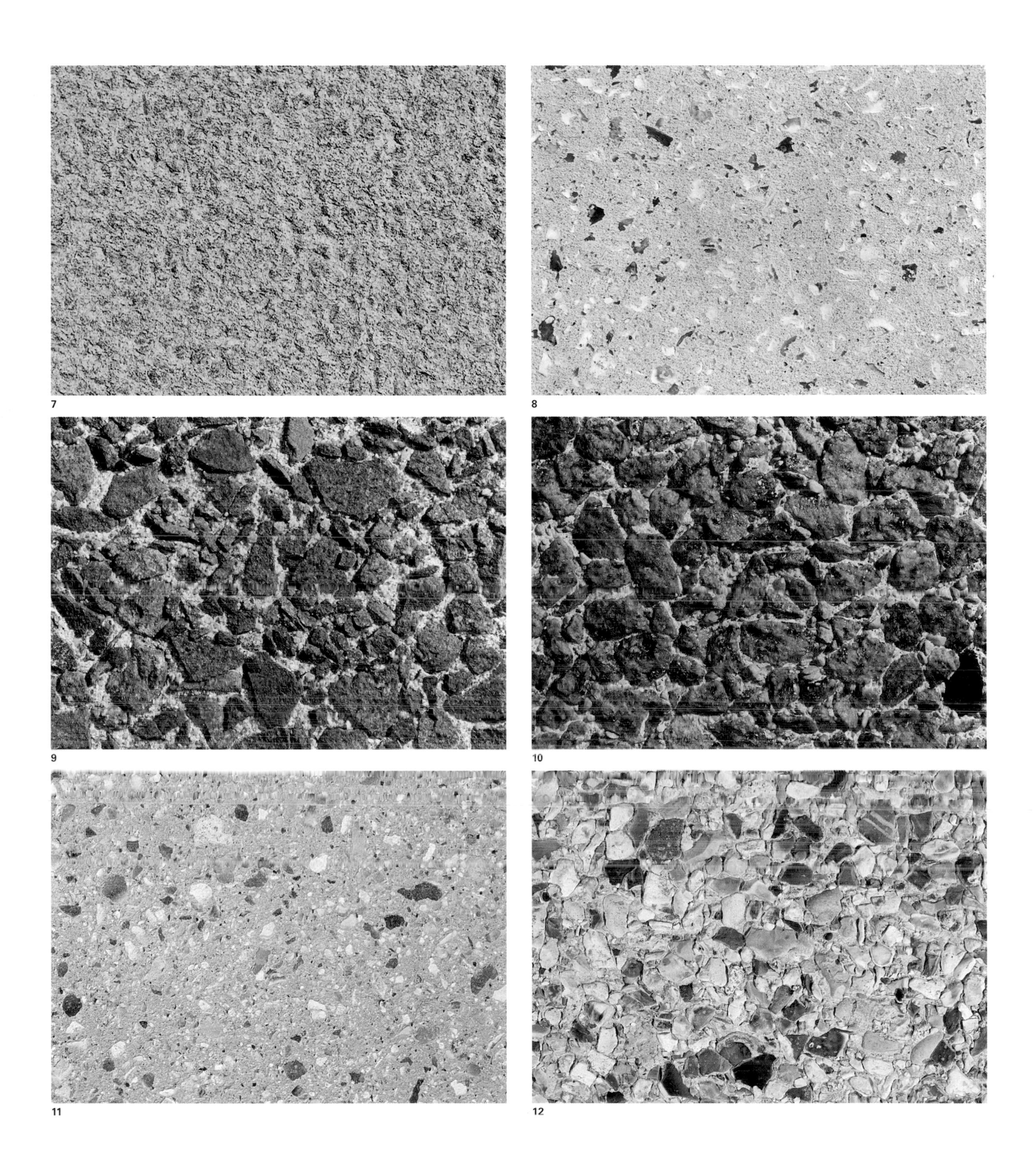

7 8 9 10 11 12

Examples of surfaces finishes

1 Bush hammered, 2 Heavy grit blast, 3 Bush hammered, 4 Aggregate transfer, 5 Heavy grit blast, 6 Light grit blast, 7 Acid etch, 8 Acid etch, 9 Heavy grit blast, 10 Heavy grit blast, 11 Medium grit blast, 12 Aggregate transfer and acid etch

SPÆNCOM VISITOR CENTRE AND MAIN OFFICE, HOBROVES, AALBORG

CF Møller Architects

Location

The precast factory of Spæncom is situated in the commercial hub of Hobroves, a district of Aalborg which is some 5 km south of the city centre. You will find it along the main dual carriageway leading south out of the city, amongst rows of houses with steeply pitched roofs, and near to the Tulip Sausage factory and a tidy McDonald's eatery. It is set back from the road and on arrival you are greeted by an assorted array of precast elements standing in a vast open yard, before turning to face the graceful lines of the visitor centre.

Architectural Statement

The building has been designed as an exhibition and marketing centre for precast concrete products and elements manufactured by Spæncom. It also functions as the administrative centre for the factory and has office space and conference rooms. The building is essentially the 'gateway' into the precast factory complex and therefore has been placed in a prominent position.

The main structure consists of a column supported concrete roof slab, which turns a right angle to fold down both gable ends of the building. In the space below the roof slab are a number of free-standing concrete walls that frame a mezzanine floor which creates the exhibition area and encloses the office area below. Some of these internal walls are painted in bright colours to mark different spatial relationships and end usage. The inner face of the walls is covered in acoustic plasterboard panels to muffle sound reflection and reduce echo. The mezzanine floor and ground floor office areas are covered in rolled rubber. The entrance area, corridors and passageways on the ground floor have been left as a power floated concrete slab which is subdivided into bays by metal strip inserts.

The main building façade is framed by a grillage of widely spaced columns and beams which support a glass curtail wall. In front of this there is a filigree concrete screen wall composed of close centred slender vertical precast panels. They are spaced far enough apart to offer good views from the mezzanine floor but are close enough to filter direct sunlight and reduce solar gain through the glass panels behind it.

Discussion

Anna Maria Indrio

The project was commissioned by Spæncom, who are one of the largest precast manufacturers in Denmark. They came to us because we have worked together and have a reputation for designing prestigious and innovative commercial buildings. They wanted a building that would showcase their precast products, with seminar areas where they could show a film or video of their latest developments, or train customers and specifiers on the design, installation and performance of their products. It was a requirement that the building would also function as the administrative centre for the factory, grouping key staff into one area and improving communication and team work.

Filigree precast screen

Our approach was to consider the design of the building in three ways. First we wanted to position the building in front of the factory so that it was prominent, visible and welcoming. We wanted to design a structure that branded the house style of Spæncom, encapsulating innovation and product excellence in the way it was presented. The second concept was to make the concrete appear lightweight, elegant and graceful yet functional and not structurally redundant. In Denmark it is far easier to build elegant concrete structures using in situ concrete. Precast elements tend to be rough and rugged with a low grade 'industrial' surface finish and that is why they are cheap. The challenge for us was to design the precast elements so that they were refined and aesthetically pleasing yet inexpensive to produce. We wanted to design a building that would communicate the possibilities of prefabricated concrete to the visitor at first glance.

The building was oriented towards the west to bring in lots of natural light so that upon entry, it looked and felt surprisingly transparent and open plan. We also planned the design so that when you see the building you only notice one structural material throughout its construction.

Ground floor plan

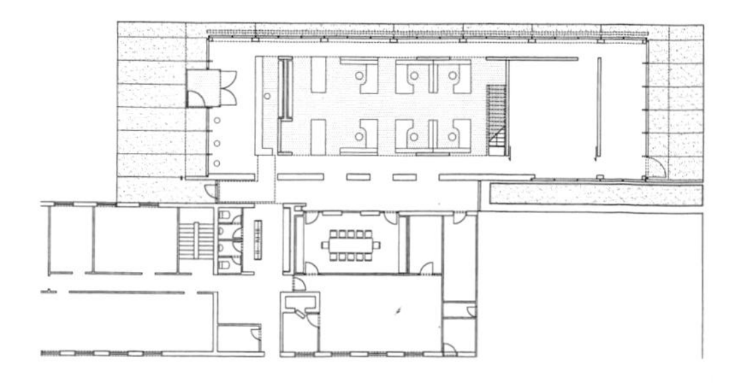

The idea of the filigree screen – the vertical brise soleil – came to me while walking the sandy beaches of Durban during one holiday. Looking up at the bright sea front apartments, the balcony walls were arranged in a very intriguing patterns creating horizontal shadow gaps like slots punched into a computer card, running the length of the building. You can see the same effect in traditional brick built grain silos of southern Italy in the countryside of my home town. The brickwork has regular slotted openings built into them to dry the grain. Both concepts were an inspiration to me. We wanted to create our screen wall structure with vertical

Internal corridor
behind screen wall

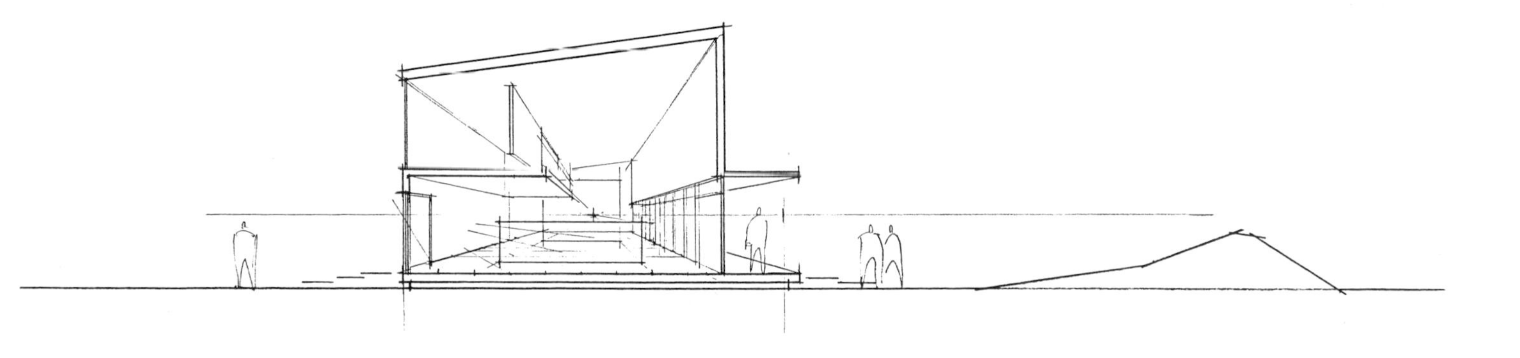

Sketch

slots that was integral with the other precast elements. The screen wall panel was detailed to be cast in 6m by 6m bays and designed to be self-supporting. The 6m high vertical blades were made as thin as possible and spaced 300mm apart, supported every 1.2m horizontally by stiffening blade beams that were the same thickness of the vertical elements. The horizontal beam elements were set back from the vertical elements so that visibly they were not the dominant visual feature. Clever detailing made the joints between each module invisible to ensure the screen wall read as one long 'trellis' 70m long by 6m high.

The precast production teams were very excited by the design concept and worked hard to develop the assembly so that it was one seamless structure. The concrete was self-finished and left as struck from the mould with no further treatment

Our third point was that inside the building we wanted to keep everything concrete as well. We divided the internal space using precast wall units which also support the mezzanine floor and create backdrops to display artwork. Colour was introduced to the internal wall faces for functional separation and as a stimulus. We did not want to use pigmented concrete as this would fade with time. Paint was much more vibrant and more expedient. Precast floor planks were used to span between the wall units for the mezzanine floor, with precast parapets forming solid balustrades along the perimeter of the open plan floor area.

Mezzanine floor

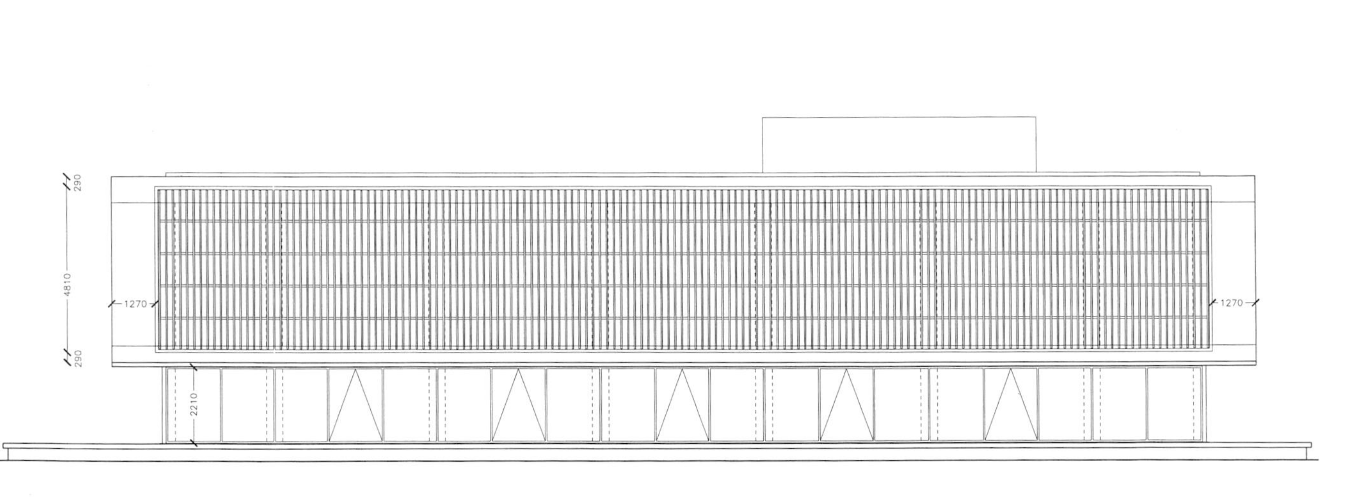

Main elevation and screen wall

Exterior view

PROJECT DATA

Client: Spæncom
Architect: CF Møller
Structural Engineer: Spæncom
Precast Manufacturer/Contractor: Spæncom
Completion: 2003

Canopy, glazing line and screen wall

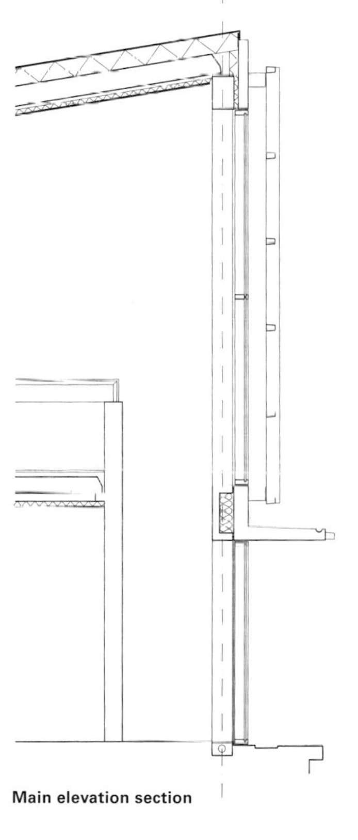

Main elevation section

GRAMMAR SCHOOL, NÆRUM

Arkitekter Dall & Lindhardtsen A/S

Painted board marked interior panels

Location

Nærum is an upper-class residential district of Copenhagen affectionately known as the whisky belt, which lies 15 km to the north of the city. By car, If you follow the highway from Copenhagen to Ellsinore, Nærum is signposted along it. The school is not far from the highway turn-off. There is no direct bus or train service from the city centre to Nærum, so you will have to change lines and bus routes to get there.

Architectural Statement

Copenhagen's new high school in Nærum for 900 pupils was the result of a design competition in 2000. The winning design compressed the school under one roof with the class rooms and study centres located along the building perimeter and around a central covered courtyard. The courtyard which doubles as an auditorium is visible and open to everyone in the building and there are flexible arrangements for group meetings, formal school functions, musical concerts and plays in this space. On three sides the courtyard is enclosed by the three-storey high school block. On the fourth, truncated side the courtyard faces the glass-walled entrance of the building.

The study areas are located in the spaces that arise in the angles between the class room wings and the glass curtain wall façade. The study rooms face the courtyard and receive daylight via skylights in the roof over the courtyard. In addition each study area has its own winter garden along the periphery of the building. The internal layout of the floor results in short connecting paths from the courtyard and recreation zones to the study centres, the quiet teaching levels and classrooms, and the library above it. It was important that the school also served the community needs outside teaching hours.

The large mono-pitch roof has a slope that follows the natural slope of the site. The sports hall is a free-standing building to one side of the main block which runs parallel to the main street of Nærum Hovegade.

Discussion

Kjeld K. Knudsen

It was difficult to find a site for the school because almost all the land in the district was developed. Any open spaces were parkland and other coveted green areas. With the co-operation of the local community the school was able to purchase a narrow strip of land occupied by a factory producing noise measurement instruments and a small group of dwellings and workshops. Even though these buildings were demolished there was never enough land for a playing field.

The design competition was run on EU lines with six architects invited to submit proposals. It was a very detailed brief with input from the teaching staff on space requirements, classroom layout and functionality. The winning design provided a building that had everything under one roof and some green space externally. Transparency and openness was critical in the design, it was a quality that the teachers most wanted for the new school.

The design allows to overview the whole school from any balcony overlooking the courtyard space. We put the school under one roof within a simple geometric form and a clearly understood internal layout. By compressing the school into one building we made space for cars, bicycles, some recreation areas and gardens on the site. Rising above the inner courtyard like a space pod, was the library building. The curving glass walled library is not a large space nor is it full of books; it is a work station for electronic communication and information with desks, computers and screens.

There are effectively two buildings on the site, the school block being the dominant one and the sports hall annexe which is alongside it. As you enter the school building you are immediately inside the covered courtyard space which rises 16m to the roof and is 50m in diameter. The classroom and study areas are located on the perimeter or the boundary of the courtyard. The building footprint is a square whose sides are 75m long, with one corner truncated to form the entrance. The total floor area is 14,000m^2 spread over three floors. On the ground floor are the administration offices, the music, art and multi-media departments which are more enclosed. On the first to third floor are the more open plan teaching rooms, laboratories and study areas.

The two sets of splayed columns painted red and rising from the courtyard support the principal roof beams that run down each side of the slotted glass opening in the roof. The columns provide sufficient lateral stiffness to stabilise the long beams. The suspended library floor is supported on two fat precast columns painted blue. The columns are hollow and have been made squat to resist sway from eccentric floor loading. The blue and the red splayed columns are quite a vivid colour and a contrast with the neutral tones of the interior walls, the ceiling and floors.

Site layout

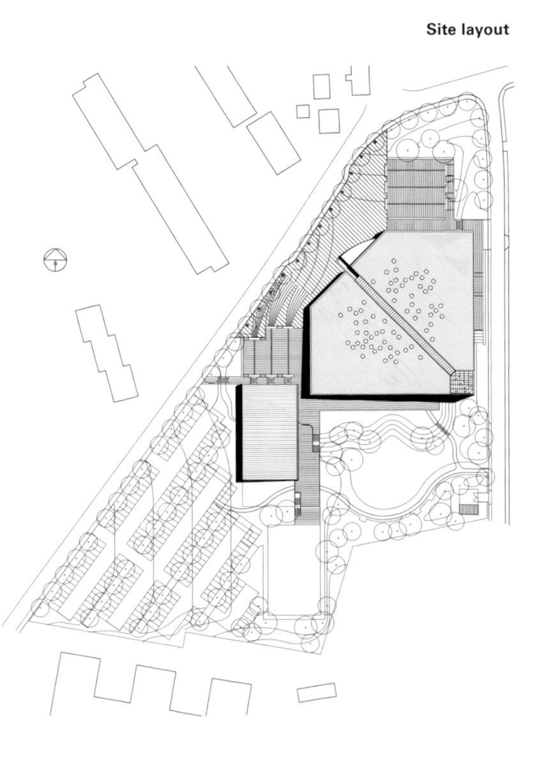

Section A-A
Section B-B

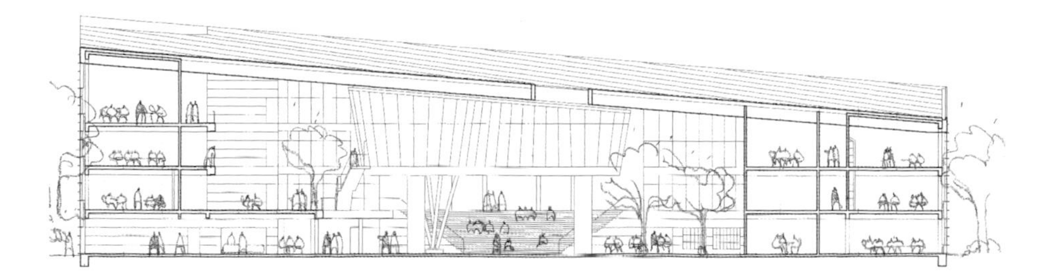

The courtyard floor is covered in a particular grey African granite and on it there is a mural designed by the artist Henning Damgaard-Sorensen. He chose different granite colours and had tiles cut into three different divisions of the square tile and then laid these in a precise geometric mosaic. The circular openings in the roof are effectively light boxes which are quite randomly arranged. Some observers may think that we have tried to depict a starry sky but that is not true, the pattern was drawn and developed without reference to any constellations. The idea was to avoid symmetry in the pattern which would then impose itself on the central space.

We chose concrete for the internal exposed load-bearing walls and the frame. On the internal precast wall the surface has been board-marked and the colour is a light natural stone. We wanted concrete as it was a very hard-wearing, durable material. The board marked panels were broken up into stone block sizes by forming grooves on the panel surface. The spacing of the horizontal grooves was in proportion to the storey height and building grid. In a way it echoes the wood grain of the pine boards on the external elevation. We chose the concrete colour based on research we carried out on the excellent concrete finish to Roskilde Town Hall which was completed in the 1960s. In the end the concrete surface had to be given a paint finish of silicate lasur much to our regret, because there were some discolouration problems.

The external façade is a glass and aluminium curtain wall system with pine wood laths inlaid above and below the window opening. The pine laths are framed between horizontal bands of aluminium fins acting as canopies to prevent rainwater from the glass soaking the timber. The wood façade is a very contemporary timber design based on traditional Nordic custom. Its lightness and warmth were chosen to harmonise with the residential buildings nearby and the pleasing environment of the school. A concrete façade would have been too brutal and a glass façade quite featureless. In recent years the Finnish timber industry has revived an old Viking custom of preserving pine by heat treating it. If you heat pine to 263° C for a set time the cells close up on the exterior to prevent the wood from ripening and rotting. The Vikings built seaworthy boats that lasted a lifetime using this method of timber preservation and waterproofing. To maintain the natural pine colour we have had it oiled, otherwise the surface will slowly turn silver grey.

Pine cladding to elevation

Mural design

Courtyard and raised library pod

The external glass is double-glazed with a coating to reduce UV penetration and heat gain. There is trickle ventilation and climate control through the windows down the central spine of the roof which can be opened or closed.

Precast Construction

Han Stig Møller, Betonelement A/S

We have a factory 30km from Copenhagen in Vibby, which means we are close to the busiest construction market in Denmark. Most of the materials that were used for precasting the panels at Nærum School were local except the sand which was brought in from Jutland. We made only a few sample mixes before we achieved the colour the architect was looking for. We were fortunate to have been given the mix constituents for the concrete at Roskilde Town Hall.

The panels for the school internal walls were cast on tilting tables on a flat bed. There were a few panels cast vertically higher up the building which did give us a few problems with an exact colour match, because the pressure against the formwork was so much greater.

All the materials were kept in enclosed silos and the concrete was batched in 1m³ lots. The panels were cast under cover in the casting sheds and vibrating tables were used to compact the concrete in the moulds. To make the board marks we used a ply sheet onto which we pinned an elastomeric formliner. This material was imprinted with a board mark pattern. To make the grooves steel strips were fixed down over the formliner and screwed into the ply. Once the concrete was compacted a machine power-floated the surface to level. The following day the panels were tilted up hydraulically and then lifted out and placed in the back of the shed where repairs to small chips are carried out before the panels are washed. They are then taken to the storage yard where they await dispatch to the site.

The panels were generally 8m long and 3m high and 200mm thick and weighed nearly 15 tonnes. Usually two panels of this size were taken by lorry to site and placed directly into position. The panels are temporarily propped until the upper precast floor planks were in place and stitched into the next lift of wall panels.

In all there were around 500 units required for Nærum which cover a surface area of 9,000m². All of them were either different in size or configured differently because of box outs, service openings, light switch positions and door requirements which had to be formed precisely. We made 33 moulds and adjusted them 300 times to achieve the final panel shape. It was not the most economic way to cast precast elements but we had a decent sized order and the continuity helped to keep costs down.

There were some problems with lime staining and efflorescence to the finished board marked panels on site. They left the factory clear but later some developed a white bloom on the surface which was difficult to remove. The contractor tried different solutions to clean it off. Acid washing failed to remove it, but as the board-marked panels had been nailed with galvanised pins a deposit of zinc was left in the concrete which turned a rusty colour during the acid washing. Heat treatment to the rust marks was tried but that failed to remove them. Sand blasting was considered but was not attempted as it would probably remove the board marking as well as the lime bloom. Moreover light sand blasting could also increase the size of the surface blow holes. In the end the contractor elected to paint all the surfaces with a silicate mineral paint matched to the exact colour of the precast panels. It also covered over some patches where the release agent had formed pale brown stains on the surface. For this reason many contractors in Denmark prefer to purchase grey precast panels and paint them to give a uniform appearance. But this could result in a long-term maintenance issue.

The vertical panels in the gymnasium were precast with a grey concrete finish and are of a very high quality. The panels were 3.8m high, 200mm thick and 8m long. They were cast in rigid free-standing metal formwork moulds with no tie bolts or corner restraints. Internal vibration was used to compact the concrete in the conventional way, which was normal procedure for a precast factory in Denmark. Acid washing removed the surface imperfections and painting with mineral paint harmonised the surface colour.

MIX CONSTITUENTS

Board mark panels
Aalborg White: 360kg/m³
Eka sand (yellow): 900kg/m³
Beach gravel (size): 1,000kg/m³
Water/cement-ratio: 0.45
Pigment: 3% of cement weight
Workability: 100mm

Top: Wide steps double as a concert platform
Bottom: First floor terrace

PROJECT DATA

Client: Copenhagen County
Architect: Dall & Lindhardtsen A/S
Main Contractor: NCC Denmark A/S
Structural Engineer: Jørgen Nielsen A/S
Precast Manufacturer: Betonelement A/S
Completion: 2004
Construction time: 15 months
Building footprint: 75m by 75m
Height: 16m
Total floor area: 14,000m²

Stairs to library pod

Looking down on the courtyard space

SID BUILDING, ÅRHUS

3XNielsen Architects

Sketch

Location

You can find the building by car, taking the E45 highway to Århus then turning left onto the Aaby Ring Road 02. The SID building is along this road between the intersection with Silkeborgvej and Viborgvej roads. There are plenty of car parking spaces adjacent to the building and there are frequent buses from the city centre to this location.

First Impression

The SID Building is a new and fresh approach to head office design. Whereas the structure of many buildings today is often concealed behind a technological, delicate sheath of glass and cladding, the SID Building derives its sophistication from its highly visible 'house of cards' in which the façade is the structure. Full-storey high, black concrete modules of various widths are piled on top of each other in a variety of patterns, alternating with windows in aluminium frames which are also full-storey height. Obviously in a game with gravity, the overall impression is one of simplicity and revelation while at the same time the variety gives the façade an interesting appearance. A box module projecting from the middle floor marks the main entrance, offset from the centre. The box modules house the main conference room and canteen. The large SID logo on the front of the box broadcasts a strong image towards Aaby Ring Road.

Grey-black pigmented panels

Architectural Statement

The building is arranged around a transverse zone, dividing the property into three main areas. In the centre are the foyer and the central atrium with the main staircase and elevator. The central area also contains the main conference room with several classrooms and smaller conference rooms on the upper floors. The participating branches of the organisation are sited at each end of the building, on either side of the atrium. The atrium is flanked by two core units, painted in red. The core units contain toilets, kitchenettes, copying facilities and cloakrooms.

The scheme went through a major revision when finalising the overall height. The building, housing several trade unions, was to have four floors and that is how the building was scaled and proportioned. But at a later stage the building owner reduced it to three floors as some of the unions were not able to move in on time. It looked squat and out of proportion when it was built but a year later the additional floor was put on.

Discussion

Jørgen Søndermark

We were very concerned with the style of the building. Should it be glass curtain walling, powder coated aluminium cladding or a more integrated structural approach and one that related to the surrounding landscape and responded to the adjacent building that was being designed. Our choice was concrete because it was an organic material derived from reconstituted rock, it can be cast easily and moulded into curved and angles shapes. Moreover concrete was the most economic choice.

We decided on a black façade to match the colour of the black metal-clad fascia of the adjacent building which was on the same plot. Our client owned both buildings and also built them.They are NCC the biggest construction company in Denmark. What we did not want to construct was a building on the same lines as our neighbour which emphasised the horizontal. Ours would accentuate the vertical.

The concept of sandwich panel construction was explored, and especially how the panels could be dovetailed, one on top of the other, to form a wall of colour with random window openings. Between the walls would span the floors, the roof and the staircase landings to create one precast monolithic composition. In Scandinavia sandwich panel construction is popular for residential buildings, and with our structural engineers we devised a programme that gave us the freedom to place panels more or less where we wanted, without compromising structural integrity. The panels stack together acting as a diaphragm wall with the floors the stabilising restraints. The load path from top to bottom flows around the window openings even if the panels above and below were stacked asymmetrically. This was a concept that we had been developing on other projects but this is the first time it has been tried on such a scale. The inner load-bearing element of the sandwich panel was cast as grey and given a paint finish.

We planned the storey-high windows openings based on four window frame sizes and then spread them in a random arrangement along each floor and over the entire elevation. We saved a lot of money on the façade because the openings where not priced. The savings we made we spent enhancing the window

Sketch elevation

Storey-high panels and window openings

design so that you could not tell the difference in frame thickness from those that opened and those that were fixed. This was quite an innovation. The only outwards feature on the flat elevation is the cantilevered seminar room hanging over the main entrance. It doubles as a canteen for staff and employees and is framed in steel and clad in aluminium.

The building is 12m wide with a central corridor and offices on both sides, some are open plan some are cellular. As you enter the building and walk into the lobby you are in a lofty light-filled atrium, with white-washed walls and a colourful staircase rising up to the third floor. The free-standing wall, four floors tall is penetrated with random window opening creating a giant collage of the sun, moon and sky and the changing light that floats by. This free-standing wall appears very slender and fragile. The wall panels are braced horizontally and tied to one another by steel channels and steel plates cast in the top and bottom of the units. These connections have been neatly hidden from view.

One of the interesting phenomena that occur with these black pigmented panels is that they will fade with time in a random way. The precast manufacturer advised us about it and we also knew this from other panels we had designed with pigmented concrete. It was intriguing that the fade would not be the same between panels although they may start off with the same colour. So as the building ages the surface subtly changes tone year by year. It is a unique character of pigmented concrete. We specified a charcoal grey, almost a black but as you can see four years on, some are still charcoal, others have faded so much that they are mid-grey. The south-east elevation which receives the highest amount of sunlight has faded more that the others. The concrete is perfectly weather-tight and sound.

We prefer to work with natural organic materials like copper, zinc, lead and wood and concrete of course because as they weather they metamorphose in colour. We do not want buildings that have a cosmetic surface that is artificial and superficially decorative with no depth of architectural integrity. They are like the glaze on a ceramic dinner plate which is clean when washed, dirty when covered in food and over time gets chipped and cracked and has to be discarded.

The aluminium window frame and the glazing panel will remain the same in colour and appearance with time. They act as counterpoints to the transforming concrete colour. The surface will also read quite differently from close quarters and such a quality is very desirable in our architecture. As you get closer you notice the same thin outline of the aluminium window frames, the lack of any visible window latches for those that open and the protrusion of the window from the building line to ensure that rain drips off the glass without soaking the concrete below it.

Office interior

Internally a black metal staircase with a wooden handrail, a red painted lift wall and white-washed perimeter walls fill the atrium and main entrance space. The lobby floor is covered in limestone flags. The colours and the internal finishes were finalised after close consultation with the building tenants who are all members of various welfare unions in Denmark. The unions historically identify with red as their corporate colour but did not want a garish, aggressive tint; they preferred a softer tone that conveyed calm and assurance.

The suspended office floors are precast hollow core planks spanning from the perimeter wall to internal precast walls. The planks are screeded over with sand and cement, and the surface covered with a vinyl floor. There are secondary staircases which are the fire escapes, at each end of the building. They are precast with the treads having a terrazzo finish.

Concept sketch of atrium

On the rear elevation a terrace of white concrete steps that lead down to a narrow but long stretch of grass on which sits an odd piece of dislocated precast concrete with steps, a handrail and a landing going nowhere. Is it a speaker's dais, a platform to practice rallying speeches or a precast sample that was left behind? No one is quite sure but its intent is quite deliberate.

Precast Construction

Neils Worm, Dalton Precast, Århus

The façade panels are all load-bearing sandwich panel construction with black pigmented facing units 80mm thick, 100mm of insulation then an internal 130mm load-bearing element. The black pigmented concrete is made using white cement, white sand and black pigments and not a grey cement which would seem the obvious choice. It is easier to control the colour using white cement because it is always the same colour whereas grey cement does vary in tint over a period of time. We buy quite a lot of white sand for our factory and used one stockpile of the sand for this project to ensure that is was also a consistent colour. We added 5% pigment as a percentage of the total sand and cement content and used black coarse aggregates. Tests have shown that if you increase the pigment dosage above 6% there is no increase in colour saturation. The weigh-batching of materials has to be precise as it is the whole focus of colour control in concrete production. The weigh machine is checked and regularly calibrated.

Atrium and staircase

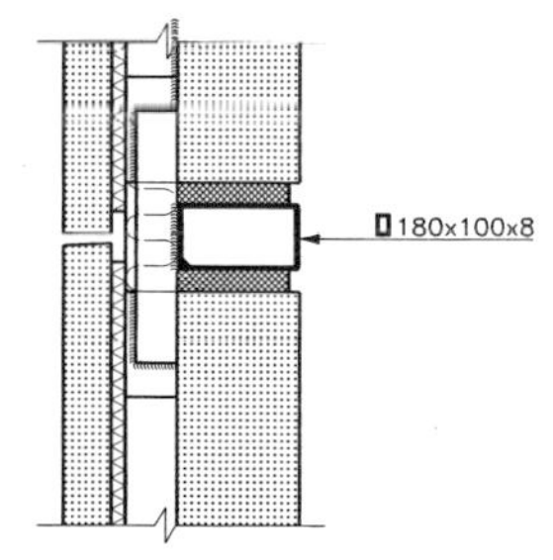

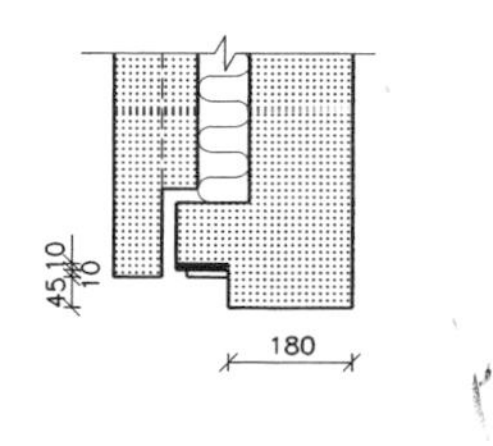

Detail: free-standing wall joint

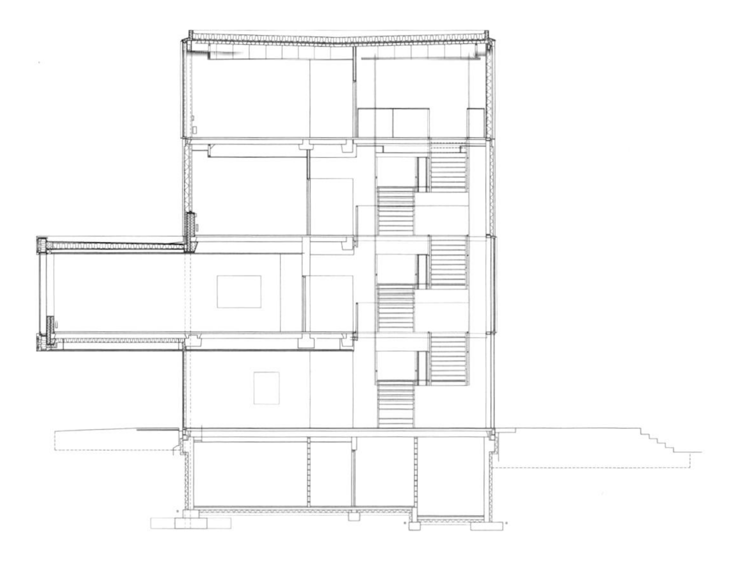

Building section

We are very concerned also about the right time to remove the panels from the moulds and to apply the acid wash to the panel. We have found that to reduce the risk of colour difference we must remove the panel from the mould by 6 AM the following day after casting. If we wait until say 11 AM to de-mould the panel the surface colour will be noticeably different and that would be unacceptable to our customers. In addition we acid wash the panels on the day we remove them from the mould and before they are taken outside to cure, in order to reduce the risk of efflorescence.

It is generally understood at least amongst precasters, that after a period of time the panels will harmonise in colour as they carbonate. It may take six months or more. Unfortunately most of our customers will not accept any difference in surface appearance of panel, no matter what assurances we give them. So we strive to keep the casting, the mould removal and acid washing to a strict regime to avoid any surface colour differences. We were blessed with a very enlightened architect on the SID Building. Not only did they understand our problems in manufacture, they actually used the subtle variation of colour tone to enhance the quality of the architecture. They were the exceptions to the rule.

Our major problem is precasting black or any pigmented concrete in winter months. We don't attempt it. During the cold season within two days of leaving the panels in the stockyard they are covered in efflorescence, which is very difficult to remove.

With all our pigmented precast panels we tell our customers that we are unable to guarantee the colour consistency because there are so many factors which we cannot control, such as external temperature, rain, sunshine and drying winds which effect surface colour. It is interesting that many more architects now prefer the subtle variations in colour tone of the panels since the SID Building was completed.

To achieve a very consistent black concrete the only way is to use single sized black aggregates and expose it on the surface. We can do this quite cheaply by retarding the concrete in the mould and then water jetting the surface to remove the cement paste to expose the coarse aggregates. This will give a textured concrete surface and not the smooth face you get with acid washing, but cheap to produce. It is an expensive operation to handle a large sandwich panel and carefully lower the face into the acid bath, then clean the surface with water without soaking the insulation and backing panel with acid or water.

The panels on the SID Building were cast in five sizes: they were all 3.5m high and either 1.2m, 1.5m, 1.8m, 2.2m or 2.7m long. There were special corner units made which formed part of the returns for both elevations to avoid a vertical joint line at the edge. The edge was given a recess using a 10mm by 10mm rebate to emphasise the corner line. In all we supplied 122 units to the projects and later supplied single skin fascia panels when the fourth floor was added on a year later.

MIX CONSTITUENTS

Black Pigmented Concrete
White cement (360kg/m^3)
White sand (615kg/m^3)
Wallhanin granite 4-8mm (215kg/m^3)
Wallhanin granite 8-16mm (950kg/m^3)
Water/cement-ratio 0.40
Black pigment 5% (18kg/m^3)

Terrazzo Staircase
White cement (532kg/m^3)
Swedish marble 5-8mm (1,662kg/m^3)
Water (242kg/m^3)

Erection of corner panel

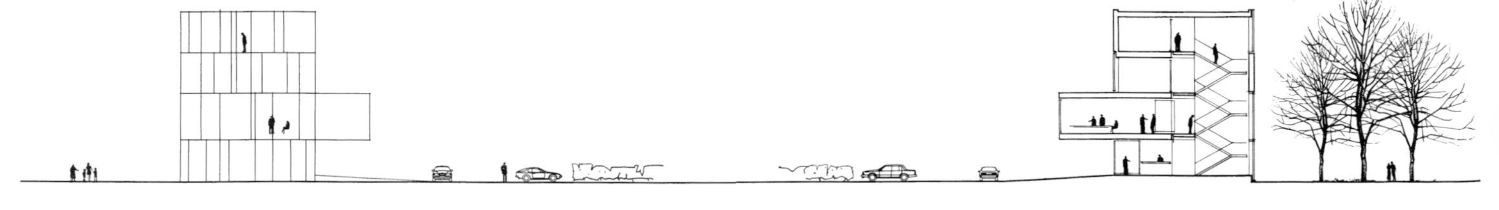

Concept sketches

PROJECT DATA

Client: SID Århus
Architect: 3XNielsen
Structural Engineer: COWI
Services Engineer: COWI
Contractor: NCC Construction Denmark
Precast Manufacturer: Dalton Ltd
Completion: 2000
Floor Area: 2,600m^2

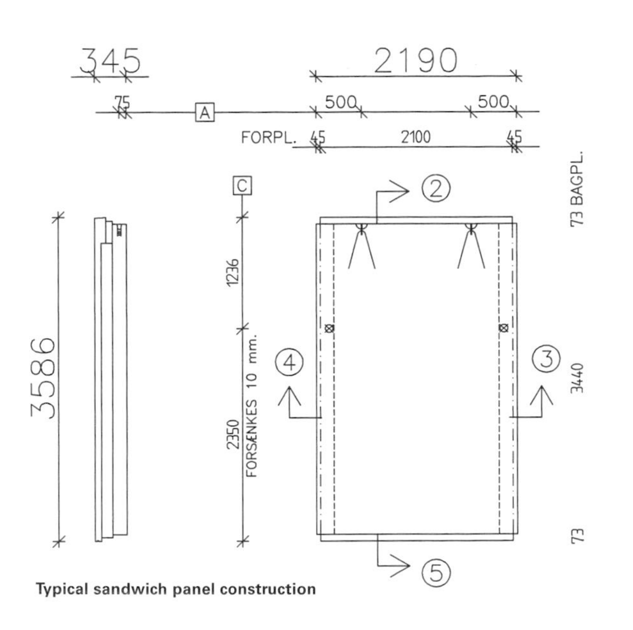

Typical sandwich panel construction

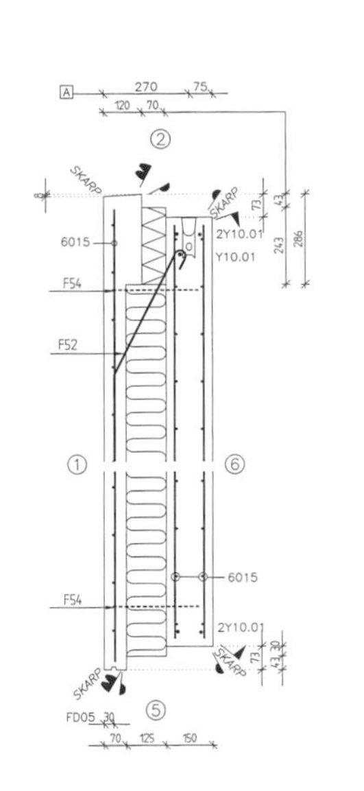

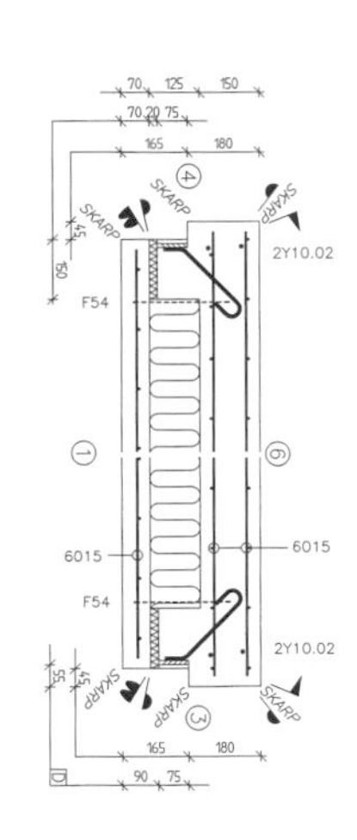

Upper: Interior floor construction
Lower: Rear and end elevation perspective

Front elevation

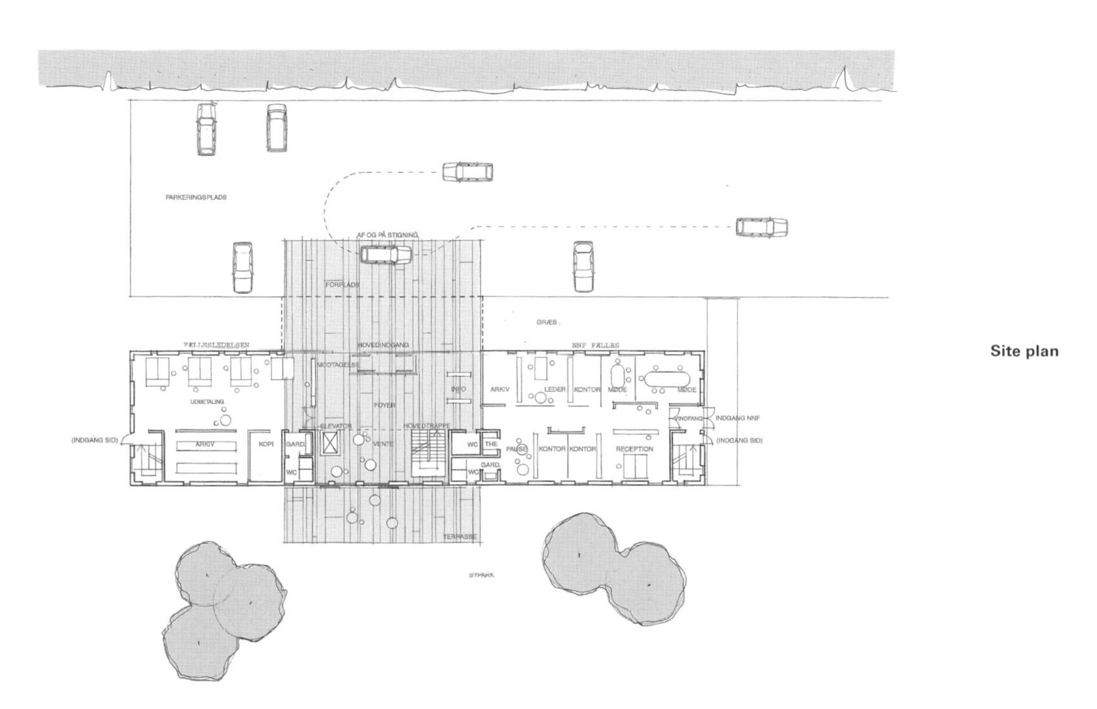

Site plan

UNITED EXHIBITS GROUP HEADQUARTERS, COPENHAGEN
Kim Utzon Architects

Location
The building sits in a redevelopment zone called the North Harbour in Copenhagen and is located along a rather derelict Stritkoverg Street. It is the first building on the site, which explains the rather barren landscape. The area will become built up over the next ten years as part of a long-term regeneration programme of the Docklands.

Architectural Statement
United Exhibits Group (UEG) develop exhibitions and displays for museums all over the world. The new headquarters building combines the development and administration centre, with an adjoining production and assembly hall all on the same site. The facility consists of a four-storey glass-fronted office block with conference rooms and communal facilities. The production hall is a double-height room with a special staging on which a variety of workshops, storage facilities and work areas can be accommodated. This is where UEG tries out its exhibition designs and concepts before shipping them to museums all over the world.

Aerial view of building

The office floors are supported by two end towers which are shaped as Egyptian pylons. The towers house the secondary functions such as staircases, elevators, wet rooms and the service installations. An entrance vestibule provides access to the ground floor, which features the entrance hall, reception and meeting facilities, plus an exhibition area and a kitchen. The upper floors are a combination of open-plan and cellular office spaces. From the second floor there is access to a roof terrace and to two residential pavilions on the roof of the production hall.

On the top floor of the building, the offices and antechambers are grouped around a central circular conference room and library, whose twelve-faceted plan is designed to reflect the points of the compass. A zinc-covered dome with a band of transparent glass along its perimeter, forms the raised roof of this central space. The light effects change with the daylight and vary throughout the day from cool blue, to clear noon light to orange-red at dusk.

Discussion
Kim Utzon

Our client designs and assembles major exhibitions on historical events and ancient civilizations which he sells to museums and cultural centres. They are the largest operation of this kind in the world. When we were designing the headquarters building they were busy putting together a major exhibition on the treasures of Ancient Egypt, which will go on show in twelve museums across the USA. It is the biggest exhibition since the one about Tutinkamum in the 1970s. It was important that the new building somehow had a reference and a link with that period of history and thus Egyptian architecture became an inspiration in the building concept and symbolism of the design.

We were daunted by the prospect of trying to infuse such architecture into the new building without seeming kitsch and superficial. How could we drag massive 'entities' of ancient Egypt into a contemporary setting in Copenhagen without risk of pastiche and of trivializing the architecture? That was the challenge. We had no buildings nearby to relate to nor a set of constraints on the site to work against. The site was an open wasteland, a new development zone of 8,000 hectares with nothing built on it and this was the first building and the first mark on a new landscape.

We hit on the idea of dividing the building into two functional parts. The front would be the office building where the design ideas were conceived and the administrative hub was located. The back would be the warehouse where an entire exhibition could be assembled, viewed and inspected by clients and finalised before being packed into crates and dispatched. Also returning exhibitions could be repaired here before they are put into storage. We put the 'hands-on' construction teams in one building and ideas teams in another. Although the two buildings are very different in character they are linked by a common construction based on monumental precast wall panels that have a stone-like quality.

The inspiration for our building was the Temple of Karnack, which has a series of entrance portals of stone called pylons through which you enter the temple building. The temple leads to a series of progressively smaller chambers until you enter the sacred chamber which is just big enough for one person. We took the pylon concept and turned them 90 degrees, located them on either end of the four-storeyed office building to act as supports for the floors that span between them. The windowless pylons are vast hollow tapering columns rising above the office roof, which houses the service installations, the lift and staircases, the toilets and kitchen. You have to go up the staircase or take the lift inside the pylons to reach the upper floors. In a sense you walk or journey through the pylon to reach the modern temples of commerce – the offices.

The office floors span between the pylons and there is a glass curtain wall to the façade. The front of the building is not as transparent as we would have liked since we had to accommodate the client's

Roof level walkway above the production hall, looking towards the office building

Twin pylon towers

Penthouse and library

wishes for a circular 'penthouse' structure and library over the top central section, and an overhanging balcony on the top floor which also acts as a brise soleil. In the back we designed a large rectangular fortress building with high windows and massive walls. It is the production hall and assembly areas for the exhibitions.

Having established the space arrangement of the building we wanted to find a material for the façade that would recreate the monumental appearance of an Egyptian pylon. We built the towers using load-bearing sandwich panels with an external panel of precast granite-aggregate concrete and not the orange sandstone of the Nile valley. It is Scandinavian sandstone, the colour of the pale grey granite sands that litter so many of our beaches. The material was deposited here from erosion of the Norwegian mountains during the last Ice Age. If the Egyptians were here that's exactly what they would have chosen for their pyramids!

To achieve a stone-like finish the external precast panels were acid-etched. We tried a number of surfaces finishes – sand blasting was too heavy and water jetting after retarding the surface was still too textured. The acid etches left the surface smooth yet slightly textured just like a piece of stone that had been split, chiselled and polished. The panels were made as large as possible to reduce cost and to reduce the number of joints on the tower. The tower face sloped inwards at a constant angle on all sides, stepping inwards 240mm for every 3.5m of its vertical height. On the longest side, the precast panels were up to 14m in length and 3.5m high. They were cast with their base and top edge at the correct angle of repose. The panel had an external skin that was 80mm thick, 100mm insulation and 150mm internal load-bearing element. Each elevation of the tower was made up of a series of flat panels that reduce by 480mm in length with every 3.5m rise in height and slope inwards; and a corner unit that is 3.5m high that wraps around the corner and is exactly the same size all the way up the building. The façade is thus marked with a series of vertical lines at the corners that stagger inwards at every floor level, and evenly spaced horizontal joint lines.

We accentuated the height of the tower by placing diamond-shaped, white tiles along the edge of the vertical joints like a carpet border. They are a special ceramic tile that was used by my father on the Sydney Opera house roof. The tiles have a slightly uneven surface due to a special thick glaze and are like those used on Mosques, diffusing the light to give them a matt appearance while glistening at certain angles like mother-of-pearl.

Precast Construction

Per Bachmann, Spæncom Precast

The production was complex as we had to cast white, acid washed panels for the pyramid-shaped towers which have all oblique sides. Thus, there were no right angles, and it was a challenge during the mould-making to observe the dimensional tolerances to achieve the exact fit for assembly on site. Furthermore the architects required sharp-edged façades, which give the building its straight lines.

We cast and delivered 2,200m^2 of panels for the whole project, which also had different lengths caused by the slanting construction of the façades. For each 2.4m height on the façade of the production hall, the front plate is thickened to create a good shadow effect.

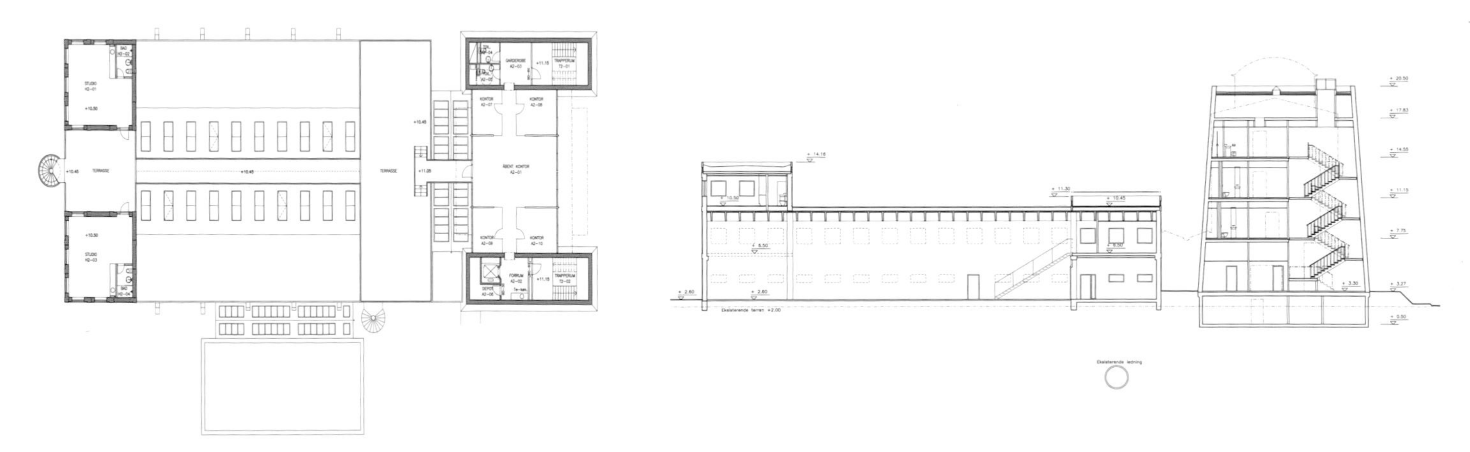

Plan at third floor level

Longitudinal section

PROJECT DATA

Client: UFN Ltd
Architect: Kim Utzon Architects
Structural Engineer: MT Hojgaard
Contractor: MT Hojgaard
Precast Manufacturer: Spæncom
Completion: 2003

Load-bearing panels and diamond tile motif

Interior of pylon tower

MAIN BUILDING, UNIVERSITY OF OULU

Virta Palaste Leinonen Architects

Location

The building is a 30 minute bus ride from Oulu airport and a ten minute taxi ride from the city centre. The University is in the Linnanmaa District to the north of Oulu. There are many buses running from the campus to the city centre, but if you are fit it is more fun to go by bicycle.

First Impression

The image of the new building in the winter snow is breathtaking. It was the image that I retained in the mind before we set off on the visit. It was a sunny day, everywhere the trees were in full leaf and the rows that were planted in front of the main building were doing their best to mask every centimetre of the cinematic outline of the building. The crisply designed surface of the exterior concrete panels, the expansive lobby with its hanging gardens on the first floor balcony, the elegant canopied walkway with its slanted steel supports, and the polished terrazzo finish to the lower wall panels in the midday sun did not disappoint. It reaffirms one's faith in the refinement and subtle beauty of a man-made rock, whose workmanship was beyond compare – yet this was nothing special according to the local precast producer.

Campus plan

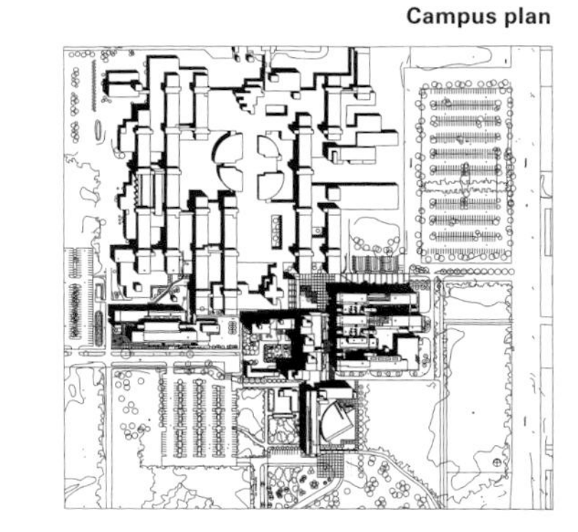

Architectural Statement

Hanna Pitkanen

All the buildings of Oulu University in Linnanmaa campus area were designed by our practice. There is a strong connection with the University going back to 1968 when the open architectural competition was held and 1973 when the first buildings were erected. The early buildings had a hard-edged industrial appearance that reflected the functions of each building and set a norm for the ethos of a college that was embryonic and striving for an identity. The buildings were aligned like beehives in regimented rows, running parallel with and at right angles to a central street that splits the campus. Each group of buildings within a development phase was symbolised with a colour that clad the plant room on the roof and enclosed the fire escapes running down both sides of the building. These escapes appear like colourful ribbons dressing the sober grey concrete façades – green for one phase, yellow for another and so on.

The objective set for the architectural competition for the main building and the administrative hub of the university, was for a more dynamic architecture that was contemporary in style and had real impact. After all it was to form the new front door of the university. In our competition-winning design we made every effort to respect the unique feature of the site and to continue the tradition of concrete-clad buildings that we had started. We placed the main building at the end of the street corridor of the campus and strove to give the building a distinctive modern character.

Discussion

The front of the building cantilevers out over the ground floor entrance, supported by splayed double-storey high columns painted black to give presence and emphasis to the building mass. The dark tone of the black columns contrasts with the light colours of the façade to create an illusion of cancelled gravity. The tall canopy structure is framed by the hard-edged blade walls of the side elevations between which is a glass curtain wall and a series of black screen slats or brise soleil that combine to express vertical movement. At night this transparent wall throws out light from within the building in all directions, illuminating the air like a vast lamp. In direct contrast, large expanses of refined pale grey precast concrete cover the bulk of the side elevations. Long slim rows of windows with long thin visors punctuate the cool grey concrete surface to accentuate the horizontal lines of the building.

Functionally the building is divided into office spaces on the upper levels and ceremonial and communal spaces on the ground floor. The main building is L-shaped in plan, and along the entrance lobby and return leg of the building the ground floor is a double-storey high atrium. As you enter the main door you walk into the lobby area and the internal street which leads to the ceremonial hall to the right, detached from the main building, and to a restaurant and conference rooms beyond. The first floor of the main building looks down over the double-height lobby space, softened with greenery and overhanging planters.

The five-storey building above the entrance has open views in all directions and contains the archives and administrative offices. On the top floor are the hospitality suites including sauna and private balcony. Over the wing section there are two floors for administrative and academic staff. The auditorium or the Great Hall is a venue for academic and public festivities, awards ceremonies, musical events, conferences and drama productions. The hall is shaped in plan like a segment of a circle, the curved face cradling the rectangular corner of the main building that opens out into the lobby. It has been designed to be totally different in scale and outward appearance from the main building, yet internally it fits into the building like a glove. The external material for the auditorium is deep blue aluminium cladding panels with black columns and black window mullions. Internally the walls and acoustic ceiling are white and the floor is

Administration building
west elevation

Covered walkway,
cranked steel supports

Administration building and auditorium from the east

oil-treated merbau parquet. Concrete roof beams span radially from the stage walls towards the back of the hall where they are supported on a ring beam.

Although the surface finishes of the main building exterior and the internal lobby walls are more sophisticated than the early buildings of the campus we have used the same palette of materials, with precast concrete being the dominant choice. We asked for a fine exposed aggregate concrete finish for the panels above the first floor with horizontal grooves indented on the surface. The grooves help to break up the flatness of the panel and emphasise the length of the building by their direction. On the ground floor both the external wall and internal lobby walls have smooth polished concrete panels which are darker in appearance as more of the dark aggregates are expressed on the surface. We preferred a neutral grey concrete with flecks of dark grey aggregates which when polished give a much darker tone to the concrete but when etched appears as a neutral pale finish.

Auditorium interior

Lobby and entrance to auditorium

There is a covered walkway from the car park to the main building entrance which runs past the auditorium. The canopy of the walkway is suspended from cranked steel stanchions painted black which are back-stayed by steel tie rods. Between the walkway and the glass façade of the auditorium was to be a pool of water but owing to budgetary constraints the area has been paved with black gravel, which creates a symbolic dry pond. Outside and to the south of the main building there is a modern bronze sculpture called 'Battery' by sculptor Veijo Ulmanen. Due to the material and colour of the bronze, it contrasts to the blue panels of auditorium and the main building in general.

The main building is in a glade in the forest in close contact with nature. The forest gives shelter and peace, a background for concentration on the scientific work. Part of the surrounding forest has been cleared to create open views, brightness and recreational opportunities. The trees have been thinned out to form areas distinctively dominated by pine or birch. The surrounding yards are planted mainly with forest vegetation and paved with slabs, cobbles and granite paving stones. Science, art and nature are living side by side in harmony.

Precast Concrete

Ilkka Kangas, Rajaville Precast Company

All the façade panels were fully detailed and engineered for precast fabrication. As a precast manufacturer we do not take on design responsibility, but we consult the architect during the design process concerning colours, details and surface finishing methods. We first make small samples the size of wall tiles in a range of colour tones of grey and with different surface finishes for the architect to finalise the colour and texture. We make the sample from mixes that we have made before, adjusting the constituents to get a close match to the required colour and tone. It did not take long for the architect to choose the sample colour and the aggregate texture for the surface finish. We then made a full-scale mock-up.

We have a number of different aggregates, cements and pigments held in silos at our precast plant. There are 32 components which is unusually high for a precast manufacturer. We keep this range of materials because we supply a wide variety of decorative finishes for architectural concrete, both as sandwich panels and single-skin fascia units. As a company we are unique in Finland and supply architectural concrete panels to all parts of the country and export them to Sweden and Norway.

For the main building the architect required a pale grey concrete matrix with a white exposed aggregate finish. The panels above first floor level – which were all sandwich construction – were given an acid-etched finish. The panels inside the lobby area and externally below first floor level were single-skin units and were wet-polished with carborundum discs to give a smooth terrazzo finish. The concrete is grey with white cement and blue-grey aggregates. The concrete surface was cut to a depth of 3mm with polishing to reveal the aggregates but without causing them to come off.

We have two concrete batching plants, one is for ordinary grey concrete and the other for batching coloured and white concrete mixes. For the concrete at Oulu we used white cement, crushed rock fines and special coarse aggregates from Finland. We did not use pigments because of the difficulty of maintaining good colour control and problems of fading with time as the surface weathers.

We batch the mixes in 750kg lots, just under 1m^3, and send it to the casting bays which are laid out on flat tilting tables in a covered shed. For the sandwich panels the exposed concrete face is cast first. We use varnished birch-faced ply laid over the steel mould to give a smooth finish to the concrete. The grooves formed in the panels are made from shaped battens of varnished plywood which are glued to the face of the birch ply base. We use silicone to make watertight seals at the groove edges to ensure there is no loss of grout during concreting as this will mar the surface appearance. We usually get about 20 uses of the birch ply surface before we have to sand it down and re-varnish it.

Upper: Exterior of auditorium
Lower: Double-height entrance corridor and lobby

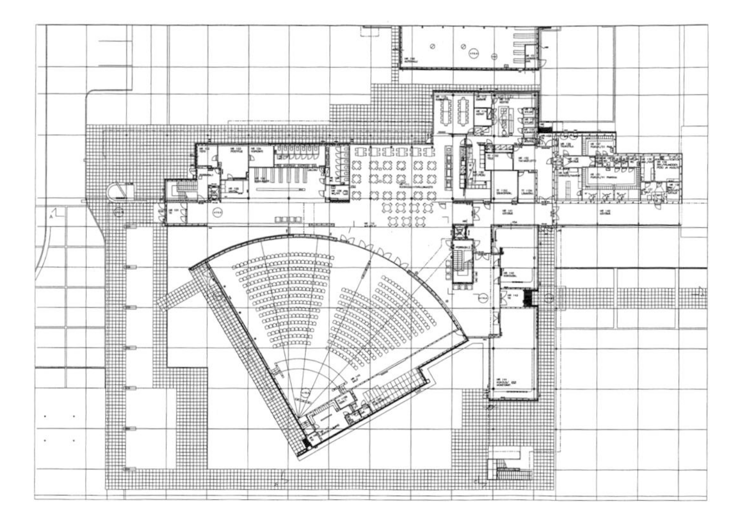

Ground floor plan

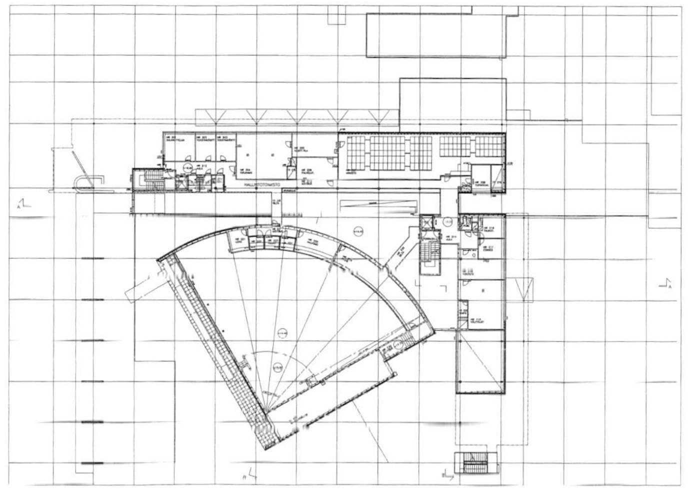

Second floor plan

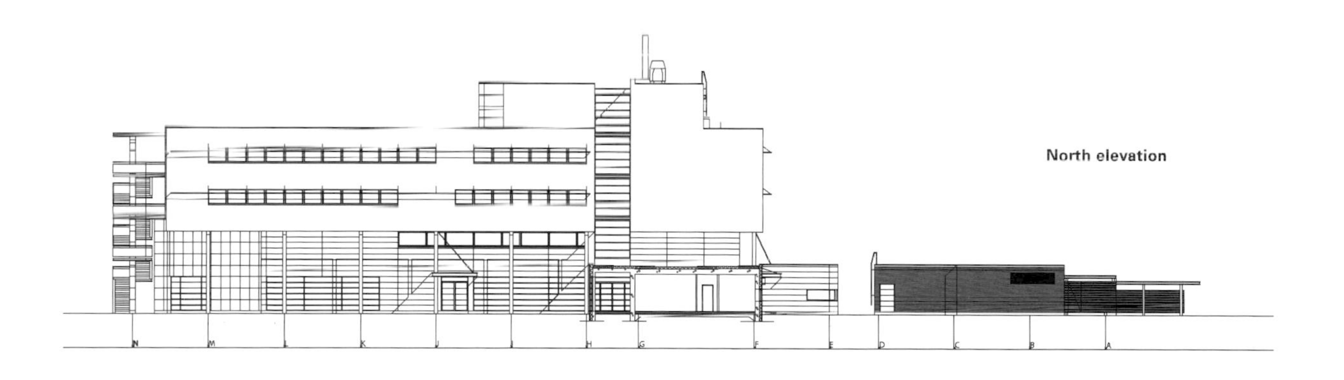

North elevation

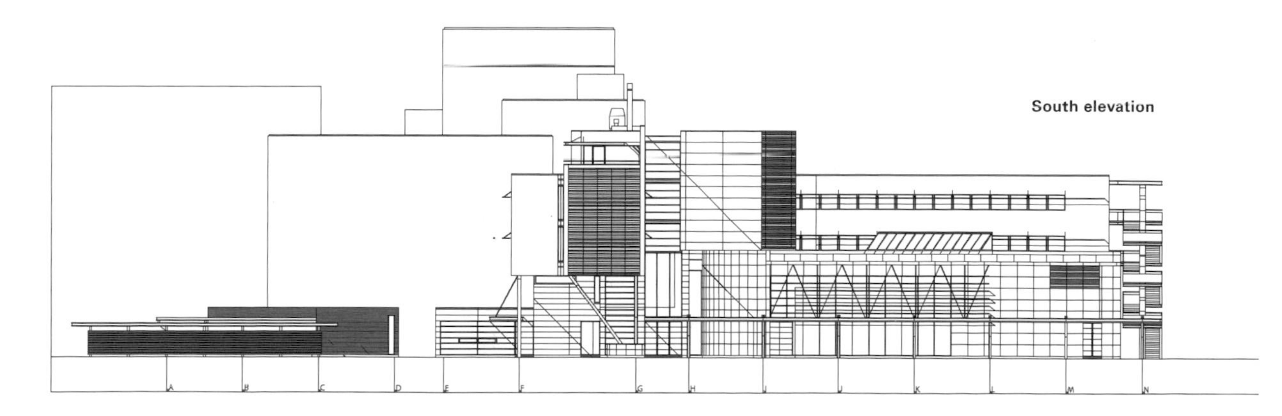

South elevation

The 70mm thick facing panel is poured and then the whole table is vibrated for two or three minutes using vibrators attached to the tables. We then insert the reinforcing ladder beams that connect the front panel to the back ones and place 150mm of rock wool insulation over the surface. The ladder beam connects the two panels in the vertical direction only, allowing lateral movement between panels. The 120mm backing concrete is then placed in the mould over the insulation and poker-vibrated. The backing concrete is an ordinary grey mix. The wet concrete surface is hand-floated to level and left to harden overnight. The following day the concrete panel is tilted up by activating hydraulic arms under the panel, until it reposes at a 60° angle. The panel is lifted out of the mould by crane without damaging the bottom edges of the slab, using chains attached to lifting eyes on the top edge of the panel. The panel is taken by overhead gantry crane to the surface preparation area.

For an etched surface to the sandwich panel, the panel face is first sprayed with water to saturate it and then immersed in an acid bath for one minute. The acid bath contains 3% hydrochloric acid. Immediately after etching the concrete surface is pressure-jetted to remove any surplus cement. This also ensures that the aggregates will not smear as the concrete dries and that no lime scale will occur.

If the panel is polished it undergoes the same acid-etch treatment and then proceeds to the polishing bay where it is laid flat and the surface is cut to a depth of 3mm using a semi-automatic wet grinding machine. The grinding machine uses various grades of carborundum and diamond cutter heads to give a fine, smooth finish. We seal the surface with special water repellent siloxane coating.

The panels are made as big as possible to reduce both cost and installation time – thus each panel is one storey high and up to 6m long. We can prefabricate them in longer lengths but a 70mm panel can curl noticeably if it is longer than 6m. The curling is outwards due to the difference in drying shrinkage between the wetter top concrete layer and denser bottom layer in the mould.

A critical feature is the joint detail. The architect wanted them to be as narrow as possible and also wanted panels as long as possible so the façade would appear monolithic. These two requirements are incompatible due to the extreme temperature range we have to design for in Finland – it ranges from -40° C to +40° C. We have to make sure the joint sealant retains its elasticity over this range. If the architect wants small joints the panels have to be smaller; thus we found a compromise with reasonably large panels and 15mm joints.

There are two types of panels, single skin elements and load-bearing sandwich panels. The single skin panels are fixed to the building structure by two load-bearing connectors at the upper part of the panel. There are also two alignment pins fitted on the lower half of the panel for positioning. The sandwich panels which stack one on top of the other to form the façade are restrained by the gravity loads from the precast floor that span onto them. Reinforcing ties connect the floor to the sandwich panel.

MIX CONSTITUENTS

Polished Terrazzo:

Single skin panels 70mm

Aalborg white cement (340kg/m³)
Black gabro fines of 0.5mm from Hyvinkää Finnland (90kg/m³)
Local grey granite 4mm (720kg/m³)
Kalanti grey granite from Norway of 4-8mm (950kg/m³)
Water/cement ratio 0.41
Air entraining agent 1.8kg/m³, air entrainment ~ 5%
Water-reducing admixture 2.8kg/m³
Flow or target slump 100mm

Acid-etched Concrete:

Sandwich panels, external skin 70mm, internal panel 120mm

Aalborg white cement (290kg/m³)
Grey cement 'Parainen' (30kg/m³)
White limestone fines from 0mm to 4mm (720kg/m³)
Pale grey limestone aggregates from Parainen 4-8mm (1,080kg/m³)
Water/cement ratio 0.44
Air entraining agent 1.8kg/m³, air entrainment ~ 5%
Water-reducing admixture 2.8kg/m³
Flow or target slump 100mm

Left: Ribbed profile
Right: Surface polishing

PROJECT DATA

Client: Valtion Kiinteistölaitos/The State Real Property Agency
Architect: Virta Palaste Leinonen Arkkitehdit
Structural Engineer: Insinööritoimisto A-insinöörit
HVAC: Hepacon
Electrical Engineer: Insinööritoimisto Tauno Nissinen
Landscape Architect: Ympäristötoimisto
Main Contractor: NCC
Precast Manufacturer: Rajaville
Completion: 1998
Total Floor Area: 5,809m²

Precast

Sandwich Panel:

Panel Sizes: 12.2 m²
Total Number: 125 pieces
Production Period: 6 months
Typical Cost/m²: 114.6 EUR

Single Panel:

Panel Sizes: 4.8m²
Total Number: 844 pieces
Production Period: 6 months
Typical Cost/m²: 90.4 EUR

Cantilevered front of main building
and splayed columns

Acid washing

Water jetty

Typical section

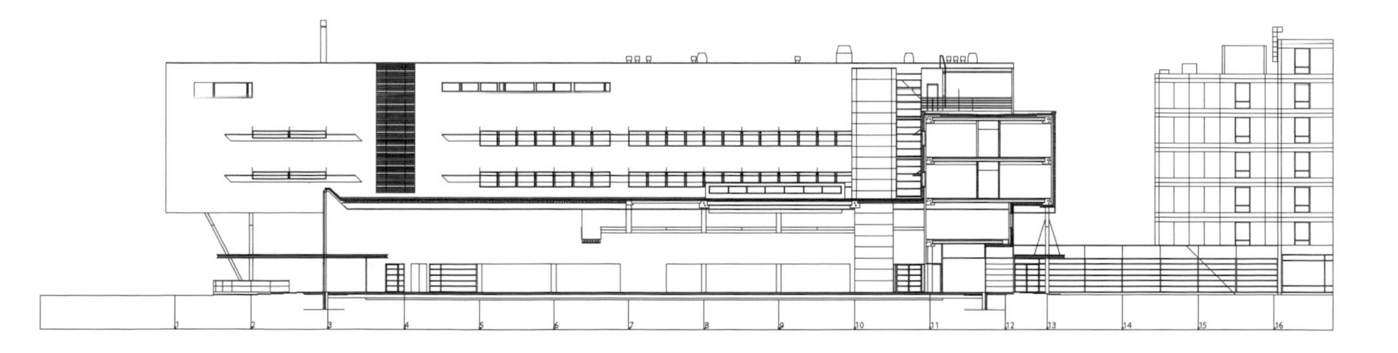

SATERINRINNE HOUSING DEVELOPMENT, HELSINKI

Brunow and Maunula Architects

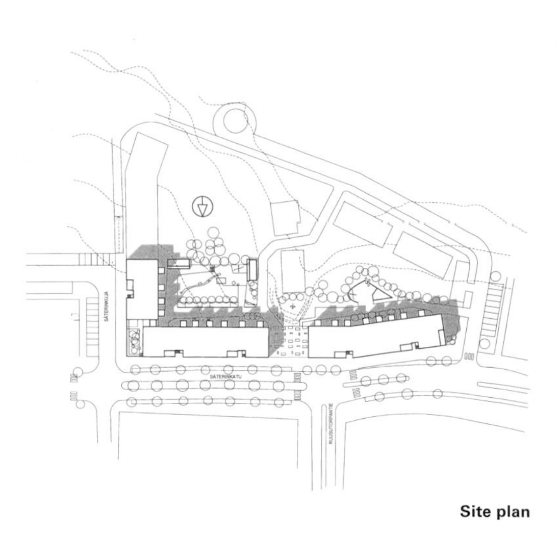

Site plan

Location

The building can be found along Saterinkatu Street, a 800m walk from Leppavaara train station, which is a 15 minute journey from Helsinki city centre.

Architectural Concept and Discussion

Anna Brunow

Fifty years ago many of the plastered façades of Helsinki houses used to be punctuated by patterns which looked like a grid of joints, perhaps in anticipation of the things to come with the introduction of precast prefabricated technology. A few decades later mass-produced concrete sandwich units flooded the markets and unfairly they were soon labelled as the cause of social and even criminal problems in the new housing areas that mushroomed in the outskirts of towns. In many European countries today sandwich units are no longer popular but in Finland they are still used in great numbers. As sensitive designers forced to work with the banality of hum-drum precast sandwich construction which contractors prefer, we would complain about the lack of scope this presents. Our office have continously been developing creative ideas and new ways to present low-cost concrete panels. At Saterinrinne we were challenged with the scheme's extraordinary use of space and a proposal for a sophisticated design solution using standard precast sandwich panels.

The buildings is on a large corner plot and fronts two roads. We were supposed to build right into the corner of the plot, although the council were planning to erect a footbridge in front of it. It was not a good idea to have apartments with their views obstructed by the footbridge and with windows that only looked to the north and east which meant that light only crept in for six months of the year. Such designs are forbidden in Finland anyway. We could put a staircase in the corner but as we already had one planned not far way it was uneconomic to double up. Staircases are expensive items of construction and must be kept to the very minimum. They are also dead spaces that cannot attract rent. So the clever thing to do was to leave the corner an open space while continuing the building line into this corner. We created a screen wall or an inside-out exterior which encloses a small courtyard garden that is open to the sky. It was a neat way to resolve this issue. We wanted to have the screen wall built in concrete just like the rest of the façade, but it proved to be too difficult. The frame was steel and infill panels were powder-coated metal. The arrangement of the screen wall shows the solid white panels as the window openings on the façade, while the gaps between them mimic the shape of the precast panels, 'reversing' the window and wall arrangement on the building façade. It was designed not to look exactly like the rest of the building, although it followed the same pattern. It was like a translucent wrapping around the building's corner. Behind the wall we have planted some trees to emphasise the reversed situation of inside and outside.

The town planners in Helsinki prefer to see more brick façades and try to enforce this onto a new development proposal. The contractor will go to arbitration to the City Council making the case about unnecessary extra cost and time delays that brickwork will impose and how this would jeopardise the project. Quite often they win – the council overruling the town planner – as happened with our scheme and that is why we were permitted to build with precast panels as originally proposed, with not a brick in sight! The contractors in Helsinki prefer to build apartments with precast sandwich panels as they are competitively priced, they are easy to erect, and the site operatives are familiar with the system. Moreover factory-finished façade panels eliminate the need for an expensive external scaffold and for wet trades associated with brickwork. But it often means we do not get building of great architectural innovation.

Screen wall corner

Because our project won in an architectural competition there was scope for new ideas and some challenging concepts. To give us the freedom to design a façade that was different and not a series of drab monolithic panels, we asked the precast manufacturer to cast a single skin façade 120mm thick which was independent of the internal load-bearing panel. They agreed on condition that we made the panels as big as possible and the same size throughout – its like hinting that you can have any colour you wish so long as its grey! What a challenge! After making many sketches and arrangements of panel shapes we hit on an orthotropic shape that was just like a large jigsaw piece with straight sides. It could work one way and if reversed or handed could work the other way to create window openings. To break the regularity of the surface and to hide the joint lines we formed broad recesses on the panel surface and these were coloured in a darker tone to pick out the irregular outline of the panels. It was such fun to design, to get the proportions right and to make the whole arrangement work technically with just

Precast 'tree' balconies on rear elevation

Saterinkatu Street frontage

Typical building section

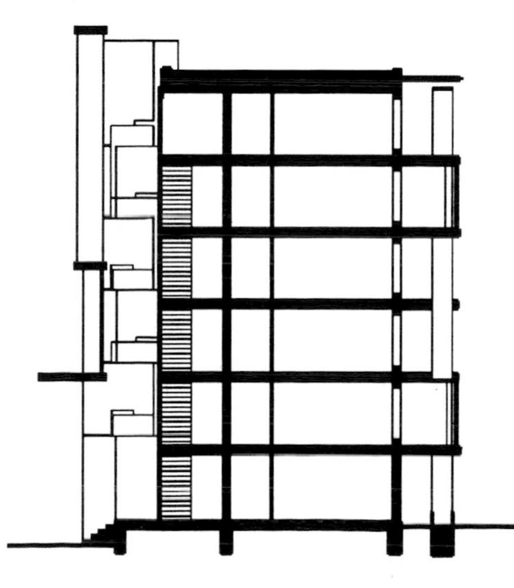

one shape. At the window openings we added a vertical sash of colour using perforated metal baffles, creating splashes of colour on the façade and heightening the window outline. From the exterior the baffles appear to be opaque and read as a solid colour, from the interior of the apartment they are a fine screen that allows light in.

Next we did the unthinkable. We persuaded the contractor to paint the façade to a carefully chosen colour scheme. The precast panels came on site as grey concrete which when it rains, darkens and develops dirt stains and looks ghastly, but this was part of the plan. The contractor could not guarantee a waterproof surface with the untreated bare concrete nor could the precaster provide it with a fine surface finish. Instead the contractor chose to apply a special water proof paint on site which was pigmented white for the surfaces and a pale green for the recessed bands in accordance with our colour plan. Where we had staircases and lift cores these were distinguished as towers that rise above the roof line with the exterior wall painted in a themed colour – yellow, red green and blue.

By contrast, standard sandwich panels are used for the rear elevations and these are painted a pale grey blue. Once again the flatness of the roof line is broken up by a series of rectilinear balcony towers advancing from the building line. The pale blue green building face is the backdrop for these symbolic 'trees' whose trunks are framed by the white edges of the balcony slab and supporting walls. At intervals up the tree towers the balcony is clad entirely of wood instead of painted white concrete or glass. These symbolise bird nests in the trees. Grey precast columns prop the balcony slabs where they cantilever from the wall panel.

As you walk between the two blocks off Saterinkatu Street you notice a layout of terracotta paving slabs interspersed with rectangular panels filled with asphalt and some with wild grasses. The architect had the idea of imprinting the windows of the gable wall on the ground. The asphalt pieces are the closed windows and the ones with grass growing is where the window is thrown open!

Precast Concrete

The sandwich panels totalled 1,650 pieces and each one was 5.1m^2 in area. The surface was cast as normal grey concrete and painted on site by the main contractor. Casting beds were made of steel and each panel weighed on average 5 tonnes.

The exterior cladding panel was compacted on table-moulds by means of an integrated shock-compacter. The interior load-bearing part of the sandwich structure wall was compacted with a poker vibrator. The product is kept inside the factory for up to two days after de-moulding, in order to ensure sufficient strength level. During that time, all fixing and minor repairs are carried out.

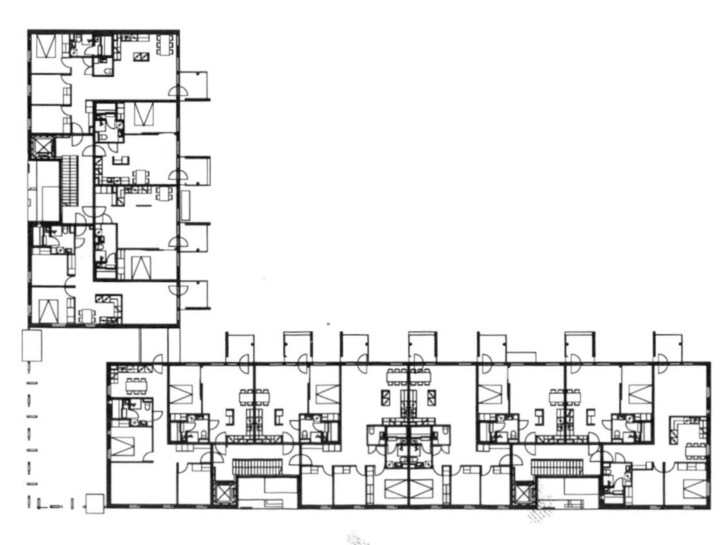

Typical floor plan

MIX CONSTITUENTS

Rapid hardening cement: 325kg/m^3
Natural filler sand: 150kg/m^3 (8%)
Aggregate sand: 0-8mm 950kg (50%)
Aggregate gravel: 8-16mm 800kg (42%)
Water/cement-ratio: 0.55
Air-entraining agent: 0.04% of cement
Superplasticizer: 1.6% of cement (Teho Parmix)

Painted front façade, L-shaped panels

Rear elevation

Gable wall with asymmetric window line

PROJECT DATA

Client: Etelä-Suomen YH-rakennuttaja
Architect: Brunow and Maunula Architects
Contractor: VO Mattila
Structural Engineer: Finnmap Consulting
Precast Manufacturer: Mikkelin Betoni
Completion: 2002
Construction Duration: 15 months
Number of apartments: 106

RASTIPUISTO APARTMENT BLOCK, HELSINKI

Helamaa and Pulkkinen Architects

Location

The Rastila Metro station is only 300 metres away and it is a 20 minute taxi ride from the city centre. Rastipuisto tronts onto a road named Retkeilijankuja and forms part of a cluster of apartment blocks in this residential district.

First impressions

The building represents a distinctive break from the traditionally painted precast façades of so many housing blocks in Helsinki. Regrettably the building is not in a prominent position so that its architecture could be admired from afar. One has to find it along a street tucked in between dull chunks of buildings. The ribbed surface of the precast concrete is crisp and precise and finely finished like chiselled stone. The attractive bands of brick slips that are cast onto the panels were a requirement of the town planners and not the choice of the architects. It is an inspiring and trend-setting housing block, designed to a tight budget and framed entirely in precast concrete, that was to encourage the city council to start trusting in the quality of naturally finished modern concrete. Timber and concrete combine well in the architecture to distinguish and define the three building blocks that make up this development.

Site plan

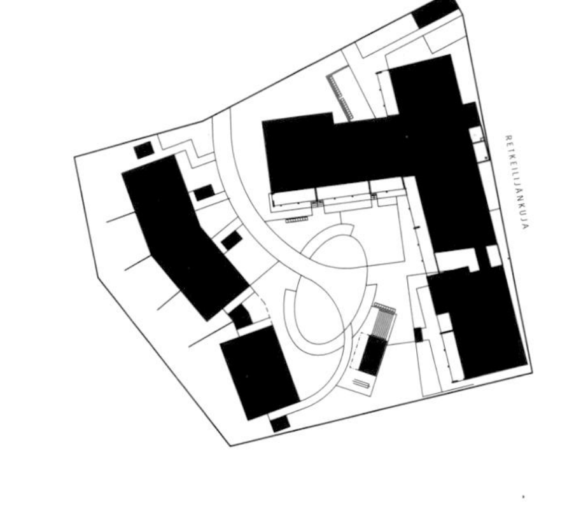

Architectural Statement

The overall plan of the scheme, the storey height, number of floors and the footprint of the building were determined by the overall town development plan. Within these guidelines the architects planned the building arrangement, the internal floor layout, the position of staircases and lifts, chose the architectural finishes of the building and the modelling to the exterior façades. The location and the environmental setting were studied before deciding upon the materials for the three buildings – concrete and timber for the main building and the three-storeyed wing, and timber for the low-rise courtyard apartments.

For the main five-storey high block that faces the street, a refined precast concrete finish in a neutral tone was initially chosen to present a more contemporary design to the public, rather than the predictable brickwork or painted look of so many apartment buildings that have emerged in recent times. The architects had to persuade the town planner that concrete instead of brickwork was an acceptable finish. They reasoned that at ground floor level a hardwearing concrete surface would be a better deterrent against vandalism and graffiti stains than soft absorbent brickwork. When the town planner conceded this point it was not too difficult for them to see the sense in continuing the precast panels for the whole building, but on the condition that half of it must be finished in brickwork.

Discussion

Jarmo Pulkkinen and Jyri Järvelä

A grey concrete finish was preferred for the precast panels to express the natural colour of concrete. A soft pastel terracotta brick was chosen for the brickwork bands that alternate with ribbed precast surface finish. Timbers came in a natural wood finish, but the exposed steel work of the balconies to the rear elevations were painted black. The grey etched finish of the precast panels was a special mix of the precast producer which had small black aggregates giving an attractive peppered surface appearance. The exterior panels are a sandwich-construction which is typical of most Scandinavian buildings. It comprises an outer-facing panel 80mm thick with the required finish, 140mm of insulation, then the inner-bearing panel, usually 150mm thick in grey concrete which acts as the load-carrying element. The internal panel face has a smooth finish and is painted as part of the room decoration. The internal and external panels are connected by a series of ladder reinforcement ties that pass through the insulation.

Ribbed finish to ground floor, brick faced above

Solar shading is provided on the rear elevations using canopied balconies to the apartments. The balcony elevations are framed with timber to scale down the hard concrete face of the building line which is inset. The three-storey wing on the east, which connects to the main block via a communal staircase, has also a precast concrete frame with a timber and brick-panelled façade with some aluminium-profiled areas. The two-storey units to the rear that are detached from the main building blocks, are all timber frame construction with balconies, in a style that is a common feature of small dwellings in Finland.

Although the plan of the main block is on a regular grid, the apartments are designed with a fair degree of flexibility. There are total of 41 two- and three-bedroom apartments within the development. Each dwelling is provided with

Brick-faced sandwich panel façade and window boxes

under floor heating, energy-saving super glass windows, a sauna, storage space in the civil defence shelter (to which every person in Finland is entitled) on the ground floor of the main building and the usual amenities. The main block is divided into discrete vertical segments, with each segment separated by a staircase or the gable walls. Within each segment, dwellings can be varied for room layout, amenity space and location of widows along the external elevation. Thus window locations along a precast panel that encloses two different apartments need not be on fixed positions, nor do they have to run in vertical alignment between floors. The storey-high precast panels stretch the full width of these vertical segments and this makes some precast units as long as 8m and keeps the number of units to a minimum, thereby reducing construction cost. Where the front elevation is punctuated by the staircases they are detailed a little differently from the repeating rhythm of the precast panels that enclose the apartment zones.

The standard design windows are of composite aluminium and hardwood construction. The exterior frame is a durable aluminium with a powder-coated finish while the interior 'warm frame' is a hardwood.

The building frame comprises perimeter load-bearing precast sandwich panels, internal precast wall panels that divide the apartments, solid 180mm and 200mm thick precast wall units that enclose the staircase and lift core and provide lateral stability. There are no columns. The floors are hollow core prestressed floor planks that span from the front to the back of the building. The building is supported on piled foundations with interconnecting ground beams to support the load-bearing wall panels. The staircases are all precast with a terrazzo surface finish.

District heating and warm water circulation with one heating element under every window is provided for the apartments. There is an air ventilation system built in with a separate ventilation machine above the apartment entrance. Therefore a false ceiling above the entrance was required in each apartment but no false ceiling in the other rooms. The floors are 200mm concrete hollow core slabs with gypsum board and oak wood parkett tile covering. The windows are triple-glazed units with a vacuum glass element to reduce thermal loss to minimum levels.

Main block street frontage (east elevation)

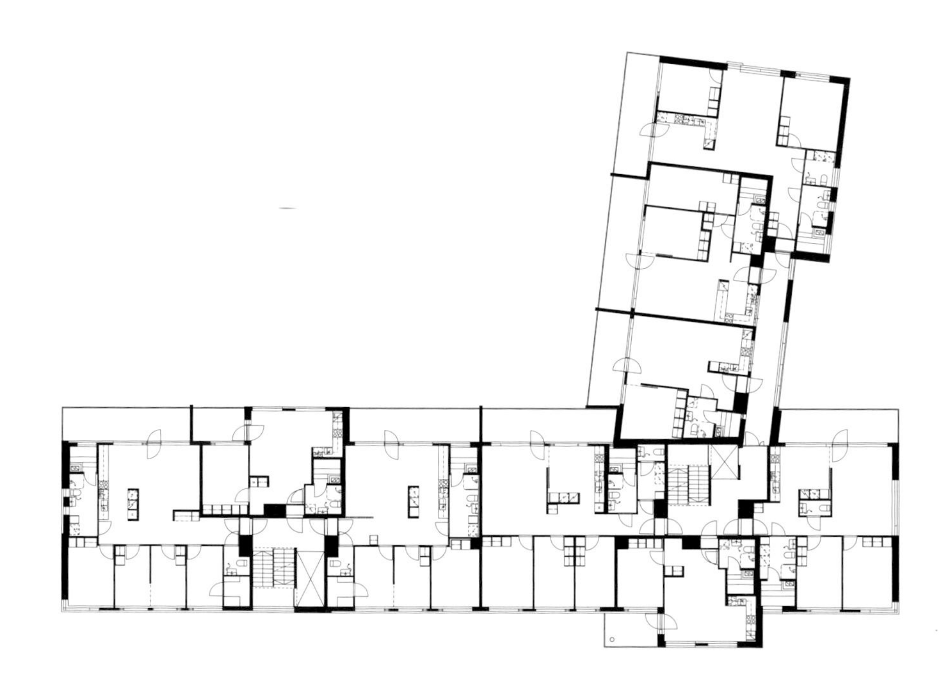

Second floor plan

Precast Concrete

Heikki Aapro, Parma Oy, Nummela

In Finland we bid for work in partnership with the main contractor. Parma is one of the biggest precast companies in Finland, and we work closely with one of the largest contractors in Finland, NCC. We generally come into the project quite early on at the design development stage and so can influence the construction and make design economies. This project is typical of many that we have worked on where every precast panel is unique because the structural size or the window openings are different. Of course the various panels will be cast from one or two master moulds that can be varied to suit window openings and adapt to the panel length.

Our factory operations and processes are highly systemised so that we make panels economically to whatever design the architects prefer. The key is to make the panels as large as possible so that the project requires only few panels. We trade in a very competitive industry in Finland as precast concrete is the dominant construction material, so we are constantly improving our production efficiency and

Rear of main block (west elevation) with timber-clad family block (south elevation) in foreground

1 Colour-pigmented precast concrete
2 Brick surface on precast concrete
3 Colour-pigmented precast concrete, surface with wave relief
4 Wooden veneer
5 Timber
6 Wooden blinds in front of bathroom windows
7 Wooden venetian blinds
8 Metal roof
9 Aluminium panels
10 Plain concrete surface
11 Balconies

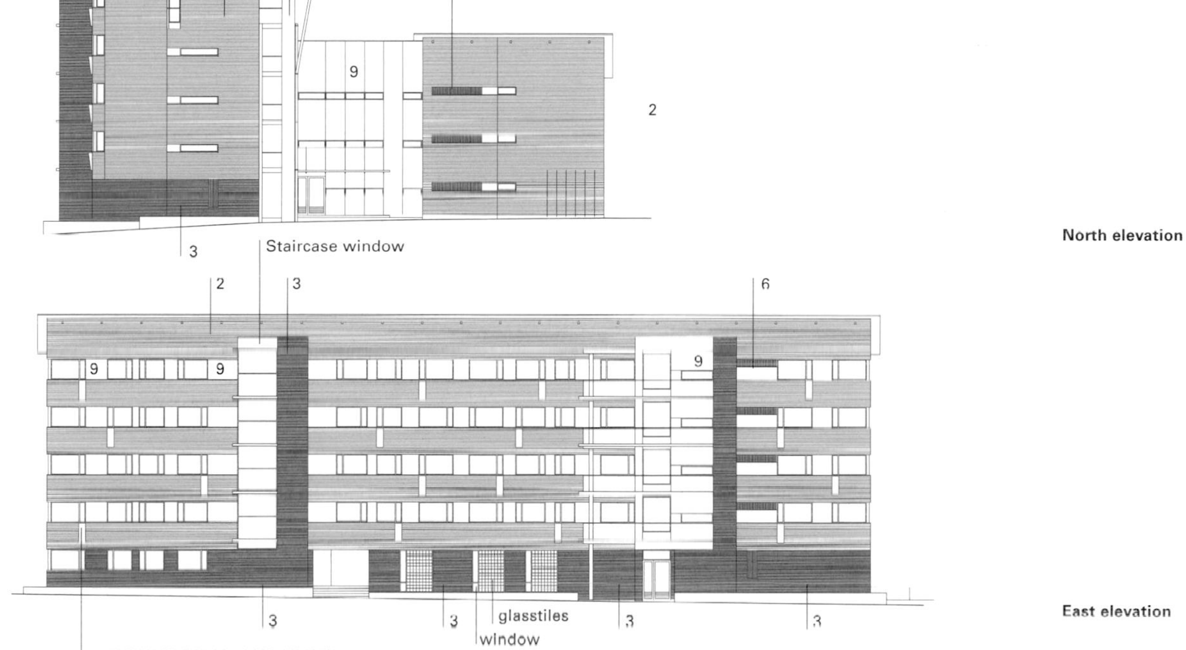

North elevation

East elevation

South elevation

West elevation

looking to innovate and save cost. To make precast concrete economic we must make panels every day in the factory to keep our unit costs down and strive to cast more than two panels from the same mould. It would be better to get more but after about five repeats from the same mould, the unit cost does not come down very much.

We use the same metal moulds 20 to 30 times before we discard them. The mould face is often lined with an HDO (High Density Phenolic Resin Overlay) plywood to give them a better casting surface, but not always. The panels are cast on long flat casting tables, with concrete fed from an overhead crane and hopper. The newly cast panel is removed from the mould the following day by tilting up the casting table hydraulically and picking up the panel by overhead crane. The panel is then taken to an area in the yard and stacked upright on railway sleepers to cure.

We place the brick slips carefully between battens that form the recessed mortar joints in the mould. To form the ridged horizontal bands for the precast face we used plastic strips fixed to the plywood. The 'face' mix – the concrete that is exposed – is placed in the mould after the mesh reinforcement for panels has been positioned and all the ladder strips for reinforcement that tie the front and back panels together, are fixed. Insulation is then placed between the ladder reinforcement over the cast concrete when the surface sheen of moisture has gone. The load-bearing reinforcement mesh is positioned over the insulation separated by plastic spacers before the backing concrete is poured.

MIX CONSTITUENTS

Cement type: rapid hardening cement 340kg/m^3
Filler aggregate: natural filler 193kg/m^3 (11% of total aggregate)
Fine/coarse aggregate: graded crushed grey granite from Kalanti,
particle size 0-2mm 264kg/m^3 (16% of total aggregate)
particle size 2-5mm 317kg/m^3 (18% of total aggregate),
particle size 5-12mm 985kg/m^3 (56% of total aggregate)
Pigment: Bayer Black from Germany, type 318, 10.2kg/m^3 (3% of cement)
Water/cement ratio: 0.57-0.59
Air-entraining agent: 0.20kg/m^3 (0.06% of cement weight)
Superplasticizer, Teho Parmix: 2.72kg/m^3 (0.8% of cement weight)

In the production of the sandwich panel for the walls, the exterior panel is compacted by an integrated shock-compacter built into the tilting table mould. The interior load-bearing element is compacted with a poker vibrator. De-moulding time is about 12-14 hours after casting.

As the concrete face is to be retarded and water-jetted to remove the surface laitance to reveal the small aggregates, the chemical retarder is brush-applied to the formwork and the plastic strips. When the concrete panel is removed from the mould it is transported to the water jetting area where it is cleaned under high pressure to remove all the retarder and unset cement paste to reveal the fine aggregate finish. By using high pressure washing it avoids any cement fines streaking over the surface and leaving a dry crusty film. The precast elements are kept inside the factory for up to two days after de-moulding, in order to ensure sufficient strength. During that time, all fixings and any minor repairs are carried out as well.

To assist the architect choosing the exact grey colour they want we supply them with concrete colour charts. We have specialists in the factory who can make small concrete samples to match our colour charts. Once the colour sample is accepted we finalise the concrete mix details. The concrete was batched in a vertical drum mixer. The concrete mixing plant is fully automatic; volume of the mixer is 1.5m^3 and the output capacity is about 20-30m^3/hour. Cement and aggregate are stored in silos with the aggregate silos heated during wintertime.

The façade panel size averaged 11m^2 in plan area with the largest unit 7,300mm long by 2,985mm high. The weight of largest panel was 10 tonnes and the total number of façade elements we supplied was 1,005 pieces. We also delivered 428 pieces hollow core slabs units for the floors.

We require six weeks lead-time from placing an order to deliver the first panels to site, but we have to know at least three months ahead when the units are needed. We receive a complete set of drawings from the architect and the structural engineer, informing us about the reinforcement required, the size of the panels to make, the thickness and where each of them is located in the building. We deliver the units to the site, and the contractor erects them and fixes them into position. This is the common practice in Scandinavian countries.

PROJECT DATA

Client: ATT/Helsingin kaupungin asuntotuotantotoimisto
Architect: Helamaa and Pulkkinen Architects
Structural Engineer: S-Planners/Ossi Rajala
Main Contractor: NCC Ltd
Precast Manufacturer: Parma Ltd
Completion: 2000

Pigmented precast ribbed
and brick-faced panels

Access to the rear community garden and courtyard

MUSTAKIVI SCHOOL AND COMMUNITY CENTRE, HELSINKI

ARK-house Architects

Location

Vuosaari is close to the sea and a 15 minute metro ride from the city centre. It is a trendy new residential suburb. The apartment blocks overlooking the sea have neat angled balconies providing wonderful views. The school is tucked behind a cluster of these bright cheerful five-storey apartments, 200m south of Vuosaari station heading towards Mustakivi Park and a good walk from the sea front.

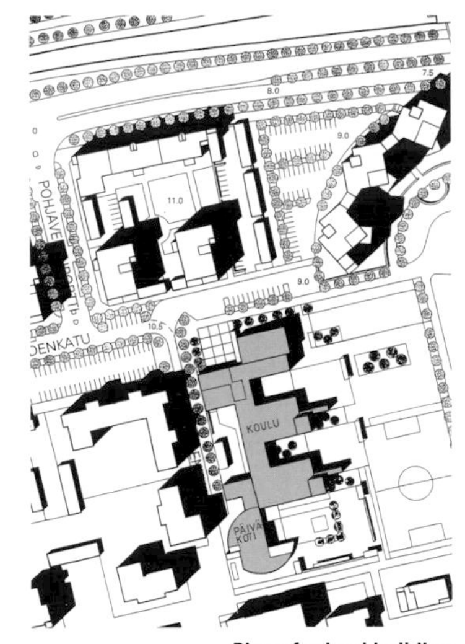

Plan of school building and neighbourhood

First Impression

It was good to have the architect guide me to the different sides of the building, as they are multi-faceted and varied. The backend of the school, where there are store rooms and the goods entrance, is a mixture of fabulous and frumpy and is not meant to be enjoyed in the same sense as the innovative and intelligent lines of the west elevations that face the playground, which you can only access by the main entrance. Here the walls are smooth, colourful and yet understated by the simplicity of their design and refined by the setting and proportions of the window details. Each window is a framed picture of the sky and fields beyond, laid on a water-coloured concrete canvas of green, terracotta, white, grey and black in a subtly changing order. The art of chemically staining precast concrete is doing well in Finland and there are sensitive artists in concrete architecture who know how to work with it.

Architectural Concept

The building has a central axis than runs north-south and stretches the length of the plot along the eastern boundary in close proximity to the street. From this central core emerge three building wings at right angles to it, facing westwards and overlooking the playground, the courtyard and open fields beyond. Two of the wings are long and rectangular and form an integral part of the central corridor building and the school. The roof line was set level to make them read as one unified block with two floors. You enter the central building at first floor level and walk along a balconied corridor with plywood-panelled class room walls leading off it to the right. The walkway at certain points becomes a bridge with staircases that go down to the ground floor. At each staircase there is a glazed curtain wall with views to the playground – here the central building becomes an atrium of natural light. As you enter the central building there is another wing immediately to the left and adjacent to the entrance. It is a large imposing sports hall with distinctive brick cladding, a sloping roof and a feature entrance. It is separated architecturally from the school to emphasise its identity. It is a public amenity to be enjoyed by the community as well as the school.

The external panels of the school block are generally of a load-bearing sandwich construction which support precasts floor planks at first floor and roof level. The supporting internal columns are in-situ concrete painted white, the edge beams over the staircase void trimming the glass curtain wall are precast elements also painted white. In the open area of the library and dining room the precast planks in the ceiling have been sprayed with warm cell insulation and painted white to give a textured fibrous finish for acoustic damping.

Administration wing

Discussion

Pentti Kareoja

Standing on the north-western corner of the site looking across the courtyard and playground towards the school and the west elevations, there are two distinctive ancillary structures rising above the roof line and protruding from the ends of the two wing buildings like aircraft steps. These are the blue-coloured emergency stairways fabricated in metal that have been extravagantly detailed with a high roof to create a screen wall to divide the courtyard and playground into small areas. As we turn to see the smooth lines of the west-facing façades, we immediately notice the interplay of solid spaces and window openings – it is a canvas of abstract concrete art, of neutral tones and regular shapes arranged in an asymmetric plan. Not all the surface colours are even, some panels are rusty in appearance, some look green with a wood stain but nonetheless the arrangement is pleasing and cathartic. Who said sandwich panels are boring! The eyes cannot settle on any one panel or resolve the rhythm of the shapes into a familiar groove,

Colour-toned wing walls and glass façade of main building enclose the formal courtyard spaces

so it fleets back and forth absorbing the colours while keeping the mind quietly stimulated. The panel lines are seamless, they appear monolithic and planar. The effect is created by setting the window line back from the panel face and making the windows storey-high so there is no vertical joint between panels.

Random colours were introduced to the exposed concrete to break up the starkness of grey and add interest to the flat elevations. The panels are arranged in chequer board fashion in no particular order – they vary in colour: you have mottled green, a rather streaky rusty brown, a mid grey, a dense black and a smooth creamy white surface. Because all the elements – windows and panels – were storey-high the windows could be placed deliberately and become the punctuation stops between the solid panel intervals. The wider windows were used for the common areas that lead into the classrooms and where children can change, or hang up their jackets and store their books. Other sizes were designed to suit the classrooms, the teachers study, the library, recreation rooms and administrative offices. The wall panels filled in the spaces; they were not a uniform length. The panel colour scheme was fixed once the jigsaw was in place.

Where the two colourful wing walls meet the glass curtain wall of the central building (where the light-filled atrium is placed), the reflections in the glass become an illusory delight. The central building melts away and has no presence, the two side buildings continue on and on, the Japanese style garden courtyard and the open texture concrete panels and grass carry through to the fields and playground that are reflected behind. What an effective device to create the illusion of space in a confined area.

At the end of each wing wall a tall glazing unit was added, breaking the roofline to signal the corner of the building. The tower above the open staircase at the gable end houses the plant room and was built in blue bricks to match the blue of the staircase to give it a strong vertical line. At the southern end of the site adjoining the central building but totally separated from it is the day centre and crèche with its own entrance. It is a curved building clad in galvanized sheeting rather like a biscuit tin with a lid.

The north wall to the main entrance is quite different from the façades of the others elevations – it is painted plywood. It was the creation of a local artist who for 'a few beers' worked with the architect and picked out a series of pastel colours from paint cards and notated them on a drawing. The contractor painted the panels according to the scheme. The fascia panels were always going to be plywood as this was the end from which the school would be extended in the future.

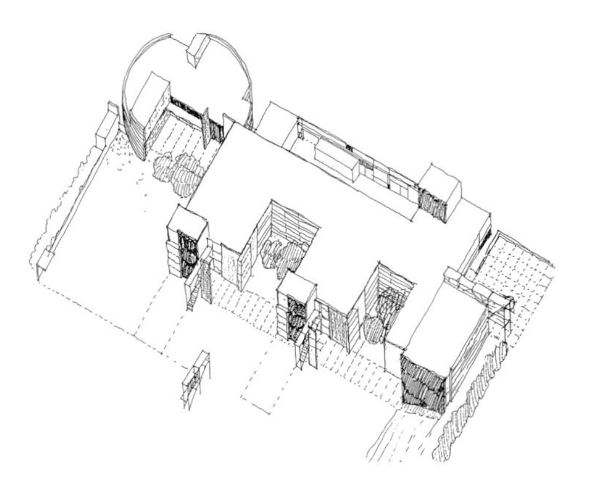

Isometric sketch of school building and day centre

Precast Concrete

Ilkka Kangas, Rajaville Precast Company, Oulu

The panels on the exterior of the school building were sandwich construction with the outer visual skin 75mm thick, then 145mm of insulation and a 160mm inner skin of load-bearing concrete.

To produce the stain effect of red-yellowish and brown panels, a white concrete panel was cast on the outer skin and allowed to harden. We then applied an iron oxide chemical stain and washed it over the surface of the panel. We applied the solution over the surface until sufficient colour saturation was achieved. The metallic irons react with the calcium hydroxide of lime in the concrete which creates a terracotta colour on the surface. It will appear streaky due to the direction of brushing over the surface and patinated due to the varying absorbency of the surface.

For the green stain copper sulphate was used instead of iron oxide and brushed over the white concrete panel as before. With an ordinary grey concrete a much darker tone to each colour would have resulted which might have looked drab. The architect required us to keep the colour tone neutral and light.

For the white panel white cement, white limestone fines and a whitish grey 0-8mm aggregate were used. There were no pigments. For the grey panel we used a standard grey concrete mix and applied a light retarder to the wet concrete surface to give it a lightly textured finish and matt appearance.

For the black panels we used a grey concrete with black rock fines, 3% black pigment and special black 0-8mm gabro aggregates from Southern Finland, Hyvinkää. We washed the lightly retarded concrete face to expose the aggregates to ensure that the surface would keep its consistent colour and not fade in time as the pigmented cement mortar colour would. We used varnished birch play in steel tilting moulds for casting the smooth concrete external panels.

Metal cladding to day centre

Playground elevation

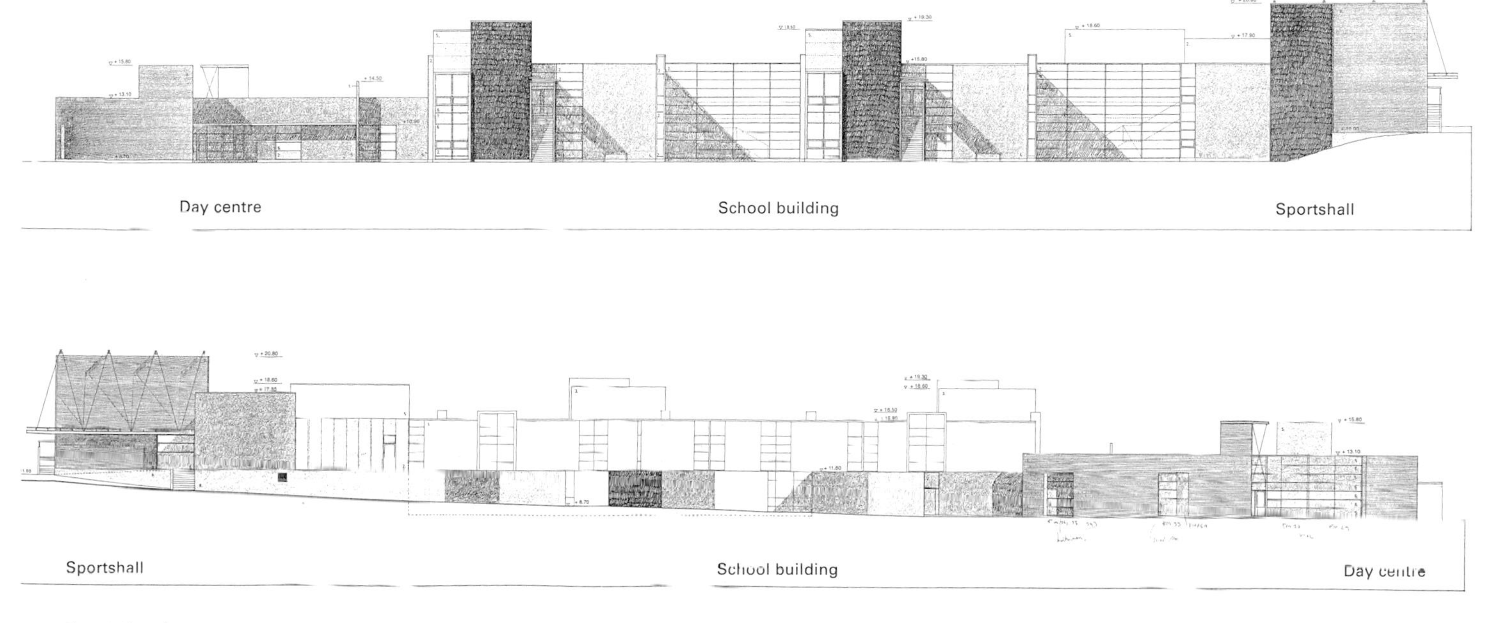

Street elevation

Central corridor and gallery

MIX CONSTITUENTS

White concrete
Aggregate: Type LK 222, maximum granular size # 8mm, Finnsementti Oy
Cement: White cement, Finnsementti Oy
Surface treatment: Washing the retarded layer

Black concrete
Aggregate: Type 35 R (black gabro of Hyvinkää), maximum granular size # 8mm, Finnsementti Oy
Cement: White cement, Finnsementti Oy
Pigment: Bayer 320, 2-4% of cement weight
Surface treatment: Washing the retarded layer

Chemically stained precast

Rear elevation, playground and stair towers

The chemical stains come to us from Bayer. Because of their very thin layer on the surface of concrete panel, their durability is fine for weather, but not so good for abrasion or mechanical wear. A few basic earth colours of iron oxides and copper sulphides are available. Surfaces can be treated with commercial sealant to avoid dirt and unwanted graffiti to penetrate into the micro-pores of the concrete. Chemical staining is something one has to experiment with by trial and error, because very much of the result is dependent on the absorbency of the surface of the concrete and its pore structure. The material cost of chemical stain is very minimal, but labour cost is the one which has the major effect, and it can add more to the cost than a standard pigmented concrete.

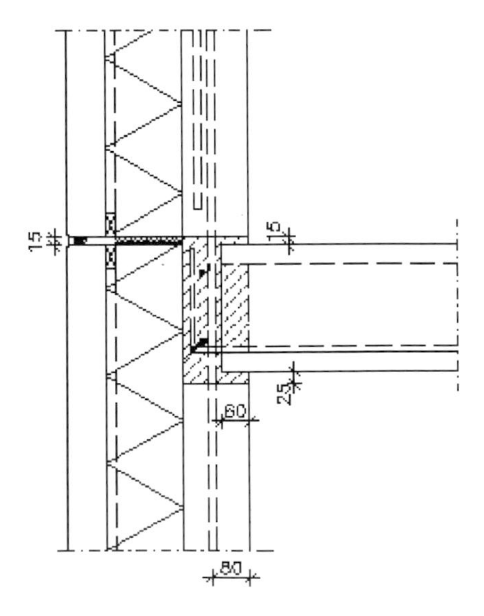

Sandwich panel and floor construction

PROJECT DATA

Client: HKR-Rakennuttaja, Helsinki
Architect: ARK-house arkkitehdit Oy
Structural Engineer: Finnmap Consulting Oy
Main Contractor: Hartela Oy
Precast Manufacturer: Rajaville Oy (Façade Panels only)
Completion: 1997
Total Floor Area: 4,965m^2
Coloured precast concrete panels: 21 pieces, 464m^2
Typical size and weight of a sandwich panel: 22m^2, 10 tonnes
Cost/m^2: 130 EUR (taxes not included)

Sketches of panel colours and windows

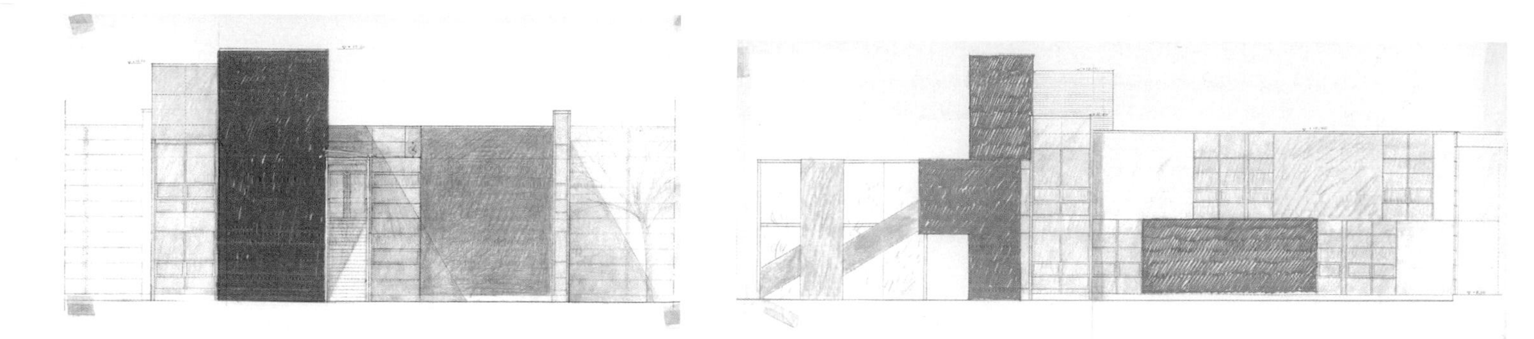

Interplay of solid panel colours and glass

9 AVENUE RENÉ COTY, PARIS

Christian Hauvette Architects

Location

The building is located at 9 Avenue René Coty, a pleasant boulevard with cafes and boutiques, in the 14th arrondissement of Paris. You cannot mistake its sharp angled corner, its modern exterior and reticulated external façade. It is near the Metro/RER interchange station Denfert Rochereau, and close to the famous lion statue by Bartoli. He was the man who designed the original Statue of Liberty in Paris, which was later enlarged and copied for New York harbour.

Architectural Statement

The building sits on a triangular parcel of land with a plan area of 600m^2, with one side bounded by the RER railway. The main body of the building above ground floor is fixed by the permitted building line of 3m from the Avenue, which is also the structural frontage, with the upper parts set back. The building is 28m high and 46m long while its width varies from 2m to 15m. The whole structure rests on a completely glazed ground floor. The floors of the offices are set perpendicular to the street line. This functional and structural organization makes it possible to address the frontage onto the street and maintain open views into the street.

The building does not hide its method of construction. The prefabricated white concrete structure forms two structural grids, each with a series of glazed openings. The first of them over four floors – first to fourth – is fixed by the height of the cornice of the adjacent building. This structural grid is defined as a reticulated arrangement of pilasters and cornices. Within them are a series of enamelled metal supports and glass panels.

This order is repeated on the two upper floors by a second element which expresses a finer structural grid set back from the one below, in an attempt to reduce the apparent height of the building. Above this are the plant rooms hidden from view. Here there is an adjoining roof top penthouse with conference rooms, director's offices and timbered balcony with view across the Paris skyline.

Schematic perspective of street elevation

Discussion

Christian Hauvette

The most interesting feature of the building has been determined by the constraints of the triangular plot on which we had to shoe-horn the building. It made for an interesting architectural statement and a structural engineering challenge. We like the natural grey of concrete as a surface because it looks like stone; it is the modern equivalent of the limestone of many old buildings in Paris. Concrete can be detailed and designed to respect the traditional style and rigour of the building lines set by these old stone buildings so familiar along Paris boulevards.

On the ground floor we have slender circular columns inset from the transparent all glass façade in keeping with tradition. Above it are a series of external beams and columns forming a monumental grillage of vierendeel frames, that project forward from the ground floor line and rise up four floors. The columns of the frame sit on a deep external precast beam at first floor level that runs the length of the building. The columns and beams of the vierendeel are all precast concrete, sharing width and surface finish. Within these frames are a series of thin aluminium blades like a radiator grill, comprising the mullions and transoms, that project from the glass panels they support. The blades diffuse direct sunlight, helping to reduce solar gain and internal temperature rise during the summer.

Rear elevation seen from the RER station

Elegant proportions of the precast frame

Crisp definition of precast structure and window grillage

Front elevation seen from Avenue René Coty

The highest beam of the vierendeel at fourth floor level respects the cornice line of the existing buildings. We continued the top of the building to match the height of the double-storey high mansard roofs of neighbouring buildings, creating an innovative filigree structure two storeys high that frames a glass curtain wall. The upper façade which is set back from the vierendeel line by 40cm, comprises a series of 8m tall slender, tapering precast elements at 1m centres supporting a glass curtain wall. It seems very lightweight and visibly fragile compared with the dominant vierendeel frames below it and that's the effect we wanted to create.

The building module for the façade and the building's vertical grid are based on the golden section and a derivative of a 16cm module. The floor height

Left: Corner view of the building
Right: Façade beams of lower floor

is 20 times 16cm, i.e. 3.2m. The width of the vierendeel beams and columns are 64cm and all the other dimensions and intervals are multiples of 16cm. Horizontally the vierendeel columns are 6m apart, while the upper slender ones are 1m apart. The vierendeel frame is a major support structure for the building over four floors as well as the curtain wall. The architecture of this building is construction-driven, with all the elements contributing to overall stability and structural integrity.

It meant that the frame had to sequence with the façade structure which dictated that two levels had to be erected externally, before the lower floor plate could be constructed. We had many discussions with the contractor who wanted to reduce the size of the precast elements to be able to build the superstructure one floor at a time, but that would have destroyed the architectural dynamic. They wanted to cast the 8m high slender columns in two sections, but agreed in the end to cast them as one piece and brace them until the floors were connected.

All the windows were installed from inside the building. The precast H frame of the vierendeel was stitched together by in situ concrete joints hidden behind the façade. The floors were tied to the precast frames with reinforcement and in situ stitches. The finishes to the precast elements were to fully visible from inside the building; therefore joints had to be carefully designed. In the tight angled nose of the building there is hardly much space for a desk let alone an office. Instead of leaving it a dead space we designed a fire escape, a lightweight steel spiral which was visually expressive.

We always check the prefabrication drawings and the quality of the prototype panel at the factory before production can begin. We insist on this and work closely with the precaster to achieve the finished architecture. Initially the main contractor forbade us to communicate directly with their precast subcontractor. After some persuasion on our part, contact was made! But the factory visits were infrequent as they were located in Belgium like most precasters who supply Paris. One minor disappointment was the colour of the slender columns which we called 'the fringe' of the building. It was too pale a grey – we wanted a mid-grey to contrast with the light grey mid-riff of the façade.

PROJECT DATA

Client: Pitch Promotion, Paris
Structural Engineer: CERA ingénierie, Nantes
Services Engineer: CERA ingénierie, Nantes
Main Contractor: SPE, Paris
Precast Manufacturer: R-MAES, Belgium
Completion: 2004
Construction Time: 2 years
Floor Area: 2,780m^2
Building Cost: 6.4 million EUR

Construction of the upper floors

Interior towards corner point

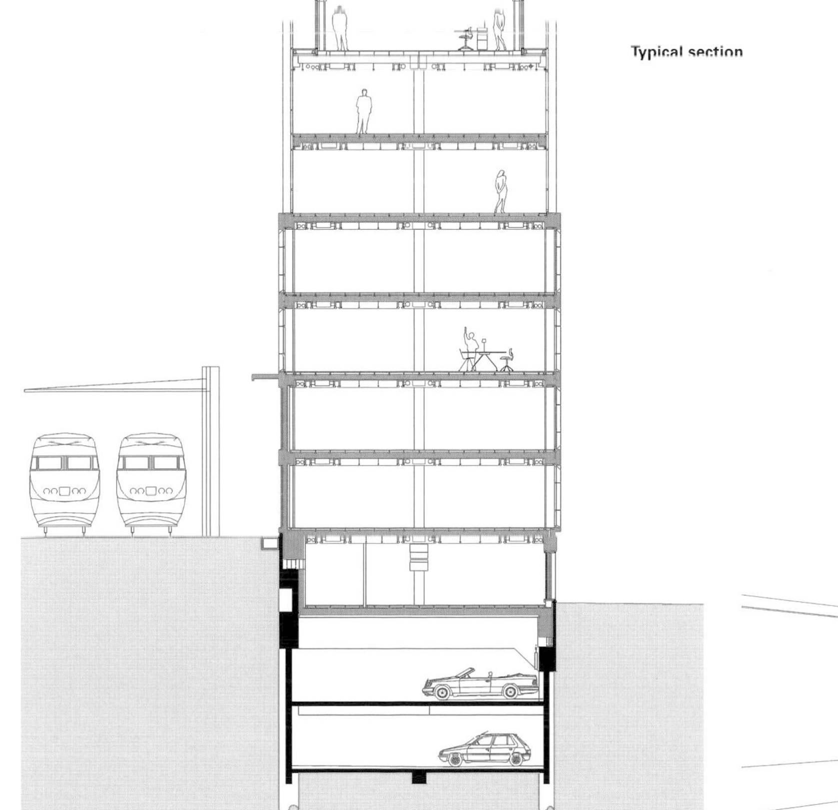

Typical section

Isometric of the triangular block

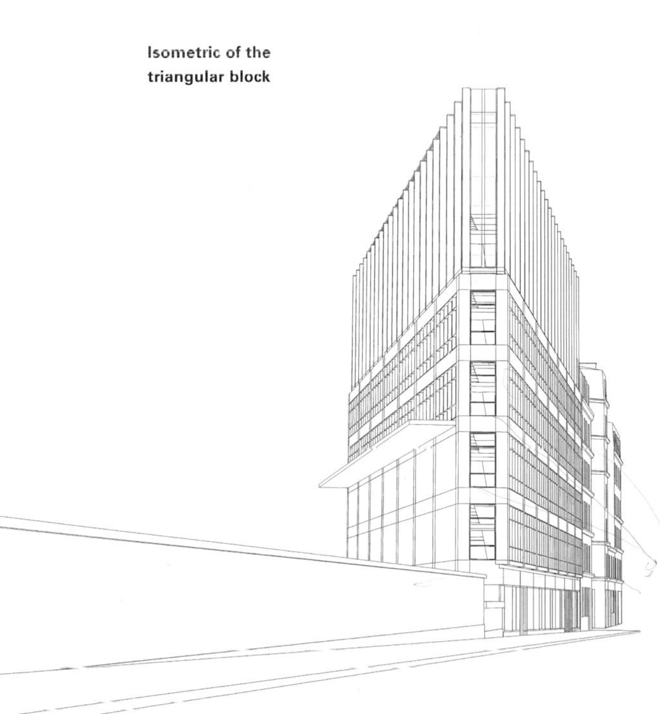

ÉCOLE MATERNELLE, PARIS

Frédéric Borel Architects

Location

The school is on one side of Cité Moskowa near boulevard Ney, between rue Leibniz and rue de la Moskowa in Paris's 18th arrondissement. Take the Metro Line 4 from the station 'Gare du Nord' until 'Porte de Clignancourt'. From the station it is a 15 minute walk to the school.

Architectural Statement

Frédéric Borel

The building is presented in the form of a monolith resting on a telluric base – a structure that tells a story. The classrooms on the upper floors connect visually with the city outside, while the ground floor is a fortress, a basal plinth supporting the upper levels and enclosing the playground with a protective embrace. We thus played hide-and-seek with the concept of enclosure and openness, introducing rooms with broad bays giving views onto the streets and public gardens, and bounding others by solid walls or small windows.

The kindergarten school is a place for learning about simple yet fundamental things in life like top and bottom, the right hand and the left, right and wrong, yesterday and today and so on. The coherence of the internal organization, the simplicity of the architecture and the homogeneity of materials must contribute to the education of the child. We avoided imagery in employing these architectural prerogatives through the pure lines of the monolith, the openness of the upper levels, and the enclosure of the ground floor. These primary forms in our view respond to the rhythm of the school day which passes unceasingly from the classroom, to the playground and back again. Considering that this building was for individuals who are not grown up, who are learning to read and write, we chose white for the wall – a sober, calming neutral colour without bias. We think that the nursery school is not an occasion to produce a great symphony of a building like a museum or a church, but rather a quieter affair like a quartet with a delicate and precise composition. We have a duty to create a sensitive architecture, whose function and form is bound to make a subconscious impression on its youthful inhabitants.

The neighbouring residential blocks fuse with and lean over the school on one side along rue Bonnet, to anchor the small island site to the continuity of the old. By its monolithic form, the school acquires a certain serenity and acts like a reference mark yet remains linked with the established buildings around it. In this modest architecture, the playful dimensions and positions of windows express the institutional character of the building. Large bay windows, letting in abundant light and small openings within the reach of a child, emphasise differences in scale.

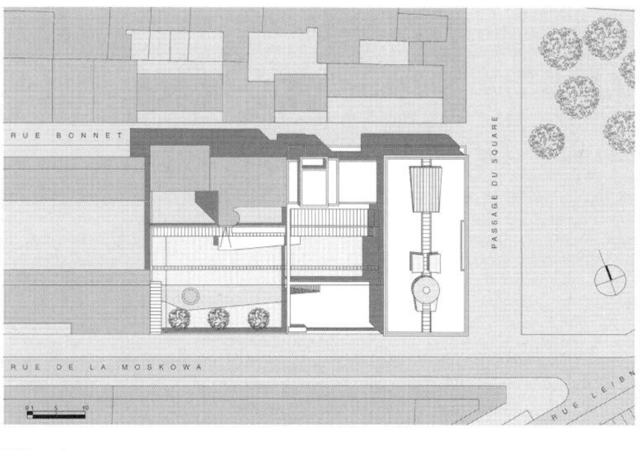

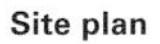

Site plan

Courtyard elevation with window box

Glass panel 'fence' abutting street (south side)

Courtyard space and activity area

Discussion

This building is a small structure that sits in the middle of a regenerated area of old apartment blocks. We chose concrete and granite as the surface material of the exterior because they were durable, robust and clean. The school has been divided architecturally into two parts: the base structure which is the ground floor and the playground, and the upper structure of two floors of classrooms that sit on the base structure. There is another small box on the upper floors along the east elevation which is the director's accommodation.

The base structure is defined as a strong box, it has a grey granite exterior and features small window openings and contains the public spaces, the staff rooms, the children's sleeping room and a leisure centre. The front elevation of the box has been modelled as a splayed wall to emphasise the smooth, hard exterior of the base. A visual separation between the base structure and the white precast box of the two upper floors that sits on it divides the children zone from the ground floor adult zone.

The school has its own character and dignity, including well proportioned light-filled interiors and a calm exterior. The intention was to create a building where children could learn about life and learn to respect the world they live in. The surfaces, the openness, the clarity and the harmony of the construction all contribute to this. The school has the maximum of outdoor play area for such a confined site. There is a large external courtyard and several terraces with one on the first floor roof.

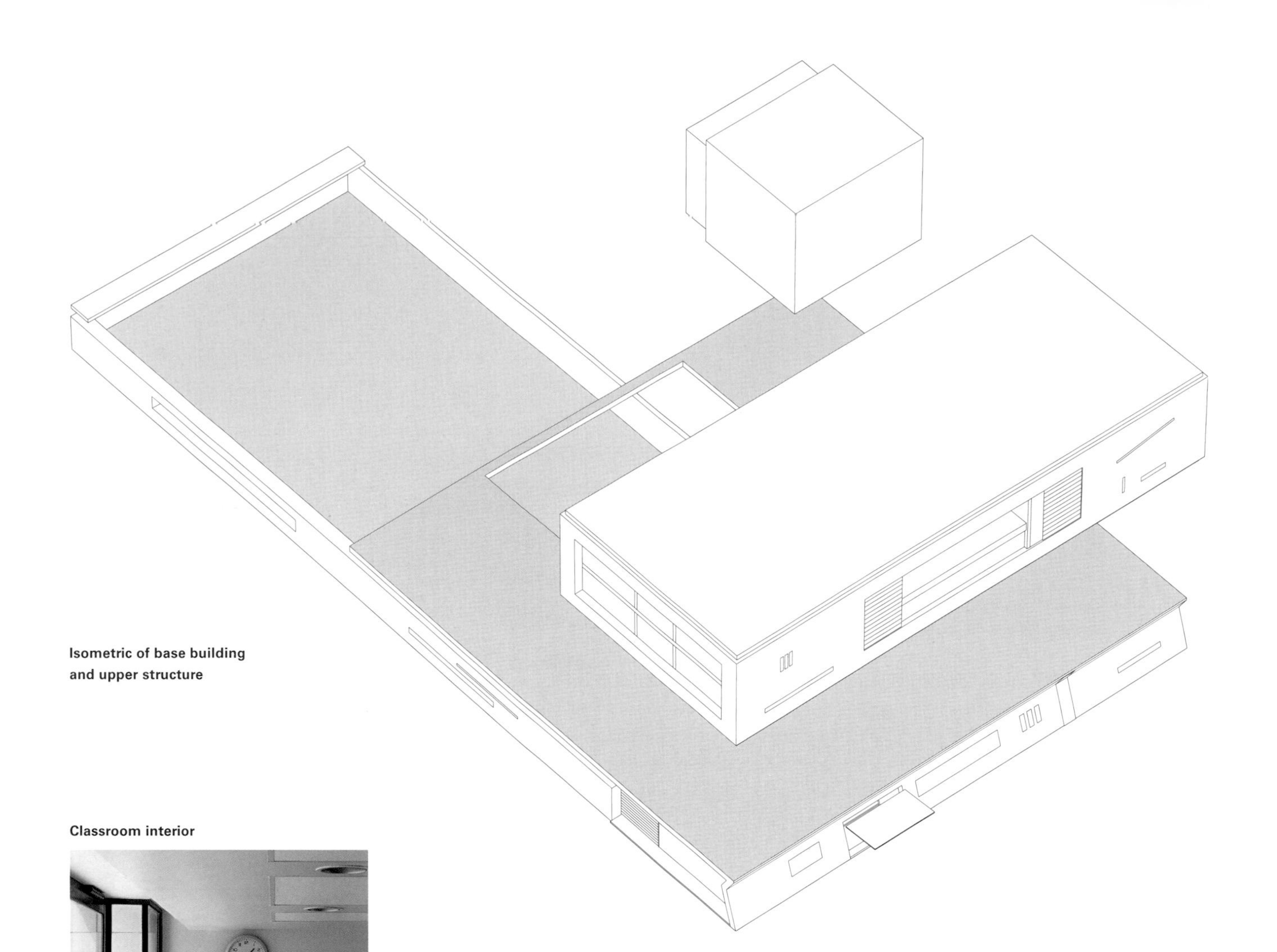

Isometric of base building and upper structure

Classroom interior

Layout of ground floor

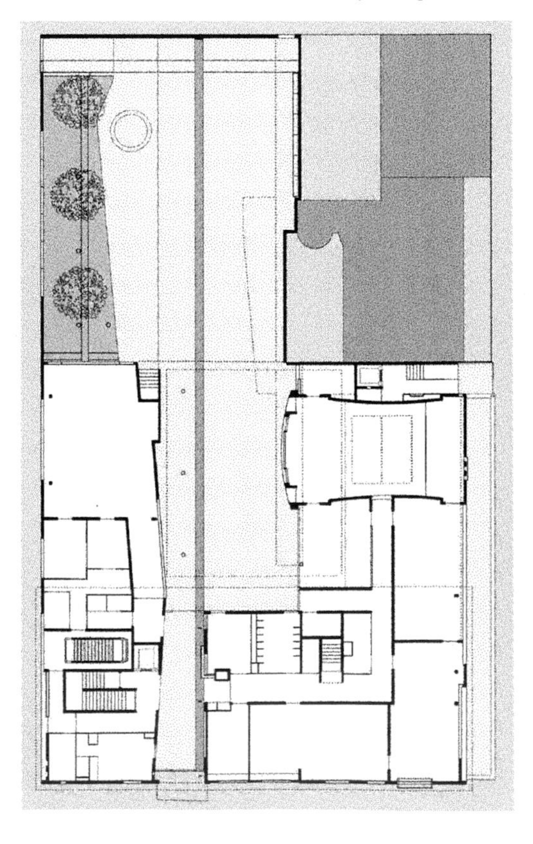

In the classrooms with their small chairs and tables, there are small windows for the children to look out of, while seated at their tables, in addition to the full height glass panels for visibility. The floor slabs of in situ concrete are covered in vinyl, the false ceiling is perforated plasterboard for acoustic damping and trickle ventilation.

On the front elevation of the white precast façade, large double height bay windows and angled lighting window slots were inserted with aluminium frames for contrast and visual stimulus. We also introduced a circular porthole window feature, a finely slatted window on the ground floor elevation and a subtle colour stripe of dark grey on the granite faced panel, to break up the plainness of the section.

There is a thematic garden along the open side of the playground in which the children can plant vegetables to see them grow – tomatoes, courgettes and French beans of course. Along the ground floor internal elevation that looks out over the school playground, there is a long window box serving as a stage for the school and community, where they can put on shows in the open air for the children in the summer. On the second floor in the principal playroom, there is a large skylight with a mobile suspended from the roof that we designed. It is an abstract butterfly made of plaster.

Principal playroom with butterfly mobile

Design sketch

Street elevation (north side) and accommodation block in foreground

Street elevation

PROJECT DATA

Client: City of Paris
Architect: Frédéric Borel Architects
Structural Engineer: S.I.B.A.T.
Acoustic Engineer: Jean-Paul Lamoureux
Contractor: Hervé
Precast Manufacturer: Morin system
Completion: 2000

Sketch of south elevation

Front elevation facing square, with granite-faced ground level and white precast upper level

A16 MOTORWAY TOLL BOOTHS, PICARDY
Manuelle Gautrand Architects

Location
Picardy is the cradle of gothic art with six splendid cathedrals and numerous churches and abbeys and the home of the stained glass industry. At Amiens there is the largest gothic cathedral in Europe, and the towns of Beauvais, Laon, Senlis, Noyon and Soissons have their massive cathedrals and beautiful old town centres. The five toll booths of the motorway are located between Abbeville and Boulogne on the feeder roads that lead onto the motorway through this historic region.

Architectural Statement
Manuelle Gautrand

I think that the toll stations on the motorways are often crudely implanted, with absolutely no relationship with the environment. I wanted this project to have strong ties to its environment and context. I was also inspired by the beautiful gothic cathedrals located in this region and their stained glass windows.

I wanted to assimilate these features to give a special character to each of the tollgates, and infuse a different spirit into each part of the expressway. The unity of the five toll stations is constituted by a thin, ethereal, slanting canopy, in a laminated, screen painted glass. Like a continuation of the natural scenery, the canopies reunite the two edges of the landscape that have been pushed apart by the 'wound' inflicted by the expressway.

They are like giant, multi-coloured screens stretched over the expressway. The support structure is introduced very discretely, being completely built above the glass surfaces, to emphasise the effect produced by the screen printed sheets.

Screen printed laminated glass

Glass canopy printed with autumn leaves and the purple-grey precast wall

The canopies pick up the characteristics of the individual region – for instance, cornfield and red poppies in Abbeville. The pictures are enlarged and superimposed.

Discussion
The design is inspired by the different landscapes along the route – the blue flowers of linseed oil cultivation, the blue of the sea, the yellow flowers of the rape seed, the red poppies in the fields, the green leaves of the trees and shrubs of the woodlands and planting on the embankments of the roadway. Another inspiration was the use of stained glass which this region is famous for. The first glass making factory was established here many years ago in the 18th century and is still the headquarters of Saint-Gobain, one of the largest glass manufacturers in Europe.

Typically, toll stations in the past have been placed on the landscape very crudely – they were seen as functional structures to contain a series of giant slot machines with automatic barriers under a metal canopy and were not perceived as architecture. Here, however, the administrative buildings and the plant rooms are not at the focus of what motorists see as they arrive to pay the toll. There is a huge glass canopy that colours the light passing through. Often the skies can be grey and gloomy and a canopy free of colour would produce the same outlook.

The glass laminate is in five colours – blue, red, yellow, pale green and mid green – one for each toll station. Each laminate panel was 1m by 1m and 10mm thick and has been toughened to withstand the impact of a man falling on them from the maintenance walkway overhead. The glass panels are supported on a framework that spans between a series of lightweight steel trusses connected to two slender steel columns fixed within the body of each toll booth island. The

trusses and columns were 2m apart. The total width of the canopy is the number of toll bays plus an overhang, while the depth is two car lengths. The toll booth cladding, the concrete plinth and tarmac below the canopy were all pigmented in the colour of the panels to continue this theme with the ground.

The administrative buildings and plant rooms are behind a screen wall to hide them from view and to create a wind break. The screen wall was precast concrete and is were pigmented in the primary colour of the canopy motif although not an exact match. The blue became a shiny dark purple-grey, the red was more a maroon, and the yellow was ochre and so on. They stretch the length of the toll area and turn right angles to corral the administrative buildings on three sides. On the 5m high precast panels branches of shrubs and trees visible in the distant fields are imprinted. The precast panels represent the landscape, and are covered with vegetation imprints that thin out towards the top.

Precast wall with imprinted plants

Close-up of precast wall

Red 'tomatoe' toilet pods

Glass conopy above toll booths

The branches and leaves were laid out on the formwork mould which was a ply construction and then fixed in place. The precaster poured a liquid latex compound into the mould and over the plants, which when set was peeled off to make the negative on which the concrete would be cast. Each precast panel was 5m high and 1.2m wide. They were butt jointed to form the screen wall and supported on a continuous ground beam foundation.

PROJECT DATA

Client: SANEF
Architect: Manuelle Gautrand
Structural Engineer: Nicolas Green, Technip TPS
Precast Manufacturer: AGP
Glass Manufacturer: Saint-Gobain Glass
Completion: 1998

Leaves and branches laid out on the formwork

HEADQUARTERS OF SOZIALVERBAND DEUTSCHLAND, BERLIN

Léon Wohlhage Wernik Architekten

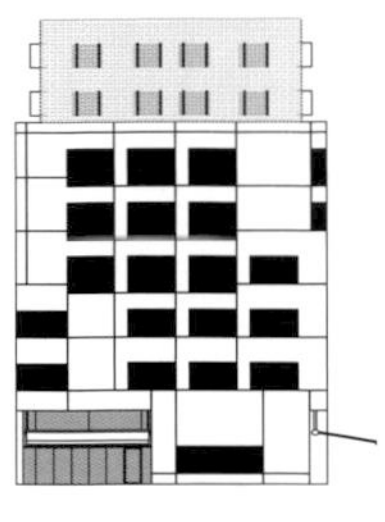

North elevation

Location

The headquarters of the Sozialverband Deutschland are adjacent to the River Spree, near the Märkisches Museum and close to the S-Bahn railway station Jannowitzbrücke which is a 10 minute journey from the centre of Berlin.

Architectural Statement

The building of the Sozialverband, an association promoting the interests of the socially disadvantaged and the handicapped, covers the site entirely and provides the maximum permissible building. It is situated in a prominent setting on the river front adjacent to an elevated railway viaduct and a busy motorway.

The office building works as a sculpture outwardly and as a place of work and abode inwardly. Protected by all-round double glazing to muffle the noise of traffic, the building offers areas for working spaces and conferences, live-in apartments and a restaurant.

The architectural intent was to give the building urban presence through a cuboid sculptural form that captures the imagination and puts up a marker for the association. The inner and outer order of the design is dominated by transparent box outs and insertions through the entire building both horizontally and vertically, admitting natural light and breaking up the solidity and mass of the hard edged monolith. Large glass areas cut deeply into the black concrete façade, while coloured window reveals act as a playful counterpoint to the solidity of the textured concrete surface.

Views of the city from lift lobbies and through the structure make up the atmospheric core of the building. The space surrounding the irregular 'greenhouse' diagonals opens up towards the entrance hall and provides ample spatial orientation. The stepped back upper residential levels and sculptural modelling give the building figurative expression and a strong personality without showiness.

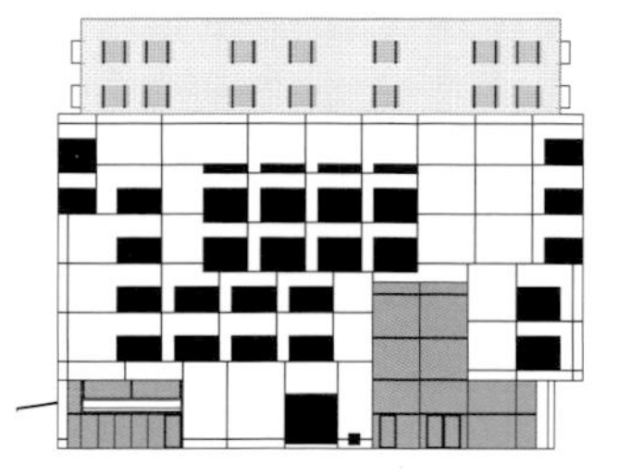

East elevation

South elevation

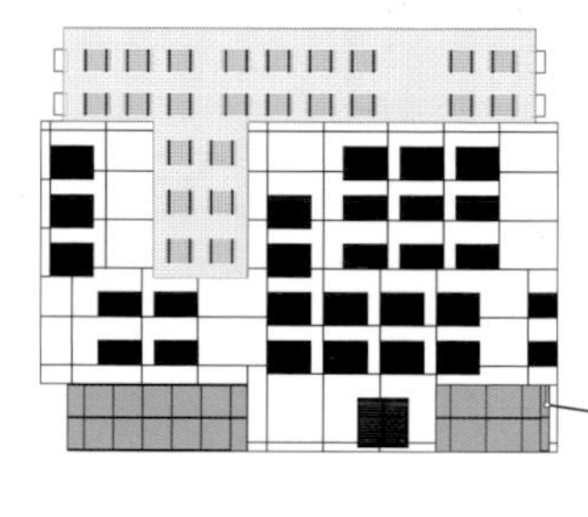

West elevation

West elevation

Architectural Discussion

Hilde Léon (LWW) and Andrew Strickland (formerly LWW, project leader)

The name of the client organisation goes back to 1918 just after the First World War when it was a political lobby group for the victims of the war – the injured, the impoverished and the orphans. As the original site of the building was east of the Berlin wall, during the Cold War years it had to be rebuilt and relocated in Bonn. After unification the site reverted back to the owners who raised the funds for the new building by selling the old plot in Bonn. It is an island site surrounded by open pavements and greenery which is unique for a new build so close to the River Spree. The town planners had designated the plot as a green space for the new apartment buildings that would mushroom around it, but they had overlooked the ownership precedent. It was a political imperative to return the site to this social democratic organisation dispossessed and banned during the Third Reich.

The building is like a shoe box but a very sophisticated one which we have cut into and shaped in different ways to express the inner structure. It has a clear organisation whose functions are easy to read from the exterior detailing: the full-height glazing to the ground floor for the restaurant and reception, the black panelled middle section for the offices and the setback apartments for the tenants on the roof.

The height of the building was fixed by the urban plan and the programmes influenced by the planning requirements for a set number of flats and apartments within the building for key workers. They were located on the upper levels. This is a little unusual as developers prefer as much commercial space as possible for a good return. In a sense that is why we have three buildings in one – the ground floor retail, the office in the middle and the upper floor apartments. We designed the landscape of the whole area well beyond the site boundary, creating tree lined boulevards and closing streets to traffic, and by doing this we gained a further 18% more space for the new building. It was give and take; the city allowed more floor area, and got a well-designed public space in return.

The structure of the building is in situ concrete, there is a central core for the lifts and the staircase which leads to the offices and private apartments. The middle and upper residential floors are divided into cellular spaces with load-bearing walls and partitions. The flat slabs span no more than 7m, and where an open area is required, as for example on the ground floor and for conference rooms in the offices, circular columns are used. The 300mm thick in situ perimeter walls mirror the external window openings and support the exterior façade panels from ground to fifth floor. Casting and installing the precast façade panels was a difficult exercise as the entire elevation of the in situ walls had to be surveyed to match cast the exterior panels to fit the as-built opening. In summary each precast panel was unique for geometry and profile. The walls of the internal corridors and lift lobbies were left as raw concrete except for a painted panel in the lobby to identify each floor with a colour – blue, red, yellow, green and orange.

An innovation in the design are the corner precast panels that wrap around the edge of the building to create a sharp line. This was not easy to precast but preferable to making the edge line with two flat panels. For these can create an ugly joint prone to waver due to tolerance and building movement. The choice of a black aggregate finish gives the façade a stylish exterior that is neither overbearing nor dull. The differing panel sizes and the cut back to reveal light-filled interiors break up the flatness of the black surfaces which in different light conditions appear shiny or matt. The window reveals have flashes of colour to add interest and delight particularly at night when the colours illuminate the windows. Bright colours are at their most striking against a dark background, and the shininess of the glass heightens the contrast to the roughcast concrete.

We have worked with the precaster Imbau on our last project where we developed a beautiful green precast concrete. So before the tenders were sent out, Imbau was invited to make samples of black concrete. After many sample mixes and finishes were made, Imbau recommended a black exposed aggregate concrete with a light acid wash which would give a consistent dense colour. Pigments were discounted as this would lead to colour variation and colour fade over time. The concrete colour was achieved using crushed fines and coarse aggregates using a black basalt type aggregate. We preferred

Exposed black aggregate finish

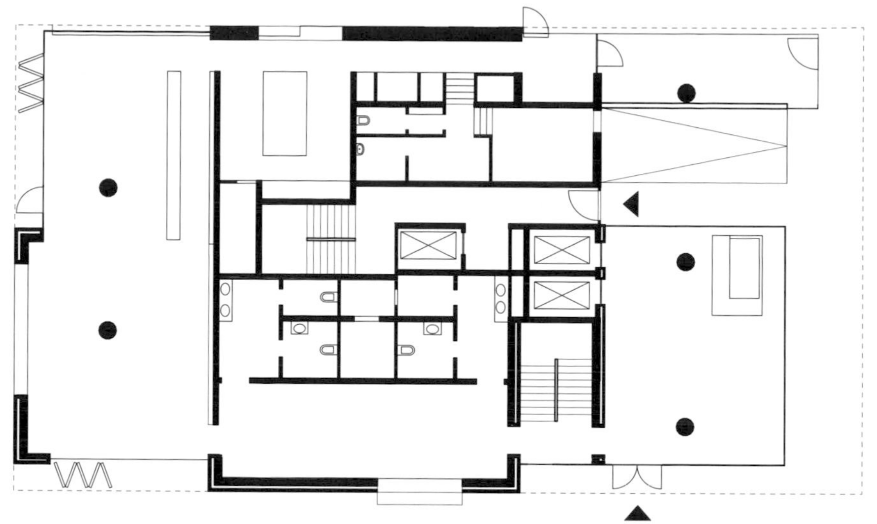

Ground floor (commercial)

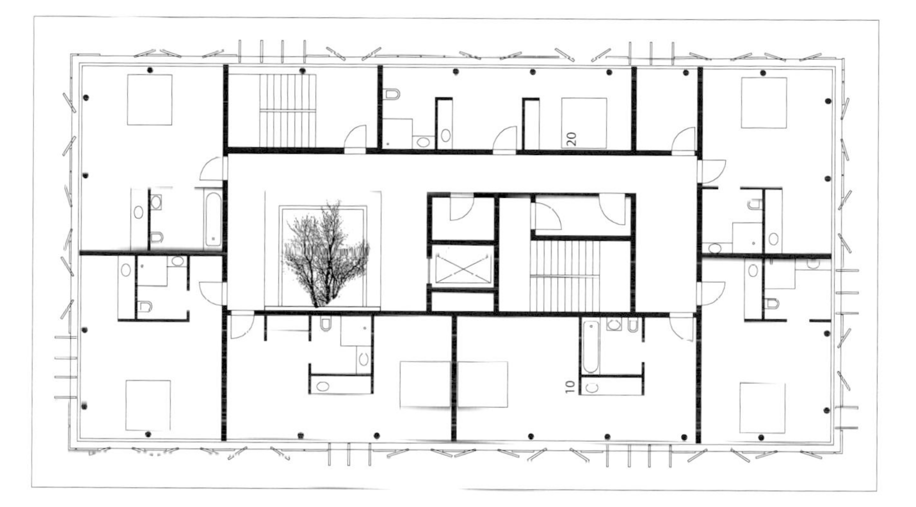

Top floor (residential)

Staircase with window reveals

the edges of the precast panels to be sharp, although the manufacturer preferred the corners to be chamfered to reduce the risk of chips and damage. Plastic protection to the edge ensured that there was no damage during transit and installation. The panels were top hung from two points towards the panel corners which give them freedom of movement.

We wanted the panels to be as large as possible and the joints as small as possible which is mutually incompatible. Small panels look just like tablets of cut stone which have no strength, no stability and are ugly in our view. Big panels are more natural in appearance. The longer the panel, however, the bigger the movement, so we agreed that the longest panel was 7m long with a height of 3.5m and joints were 2cm throughout. The larger joints emphasise the shadow gaps. The 120mm thick façade panels were separated from the in situ by 130mm of insulation.

The two-storey apartment block on the top of the building is set back from the building line and the concrete perimeter walls clad in grey colour coated metal sheeting and glazing units. We were interested in creating a building with a dominant presence in this location. Our clients accepted this idea only with some difficulty. In the end they saw it as a way to bring their organisation and its important social programme in Germany to a wider public attention, beyond its clientele of disabled and elderly people. This has worked; the Sozialverband and its Headquarters building are now known features of the Berlin landscape.

Modelling the east elevation atrium space

Reception and light well slots on the east elevation

PROJECT DATA

Client: Reichsbund Wohnungsbau- u. Siedlungsgesellschaft
Architect: Léon Wohlhage Wernik Architekten
Structural Engineer: Herbert Fink Ingenieure
Precast Manufacturer: Imbau
Completion: 2003
Gross Floor Area: 5,200m^2

Soffit above entrance

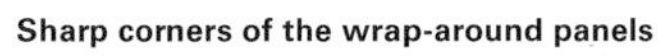

Sharp corners of the wrap-around panels

SCHARNHAUSER PARK, OSTFILDERN

Janson + Wolfrum Architekten

Location

Scharnhauser Park is near the city of Stuttgart and the airport. If you take the tram from Stuttgart it is a 20 minute journey. There is the Bus Line 122 which runs from Stuttgart Airport to Nellingen and Scharnhauser Park and takes about 20 minutes.

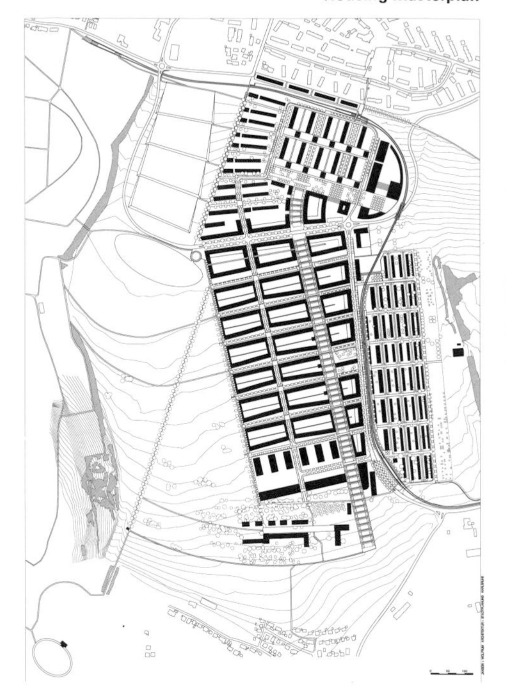

Housing masterplan

The Town Plan

Christian Holl, architecture journalist

Scharnhauser Park is a new residential development of 9,000 houses spread over 140 hectares of land, that was previously occupied by a military airbase. This new neighbourhood is part of a young local authority that was founded 30 years ago to unite four communities. This new town is being built in stages and is symbolic of a theatrical play in many ways with scene setting designs and architectural dialogue, except that it all takes place at a very slow pace. The buildings and the open spaces are the actors initiating the dialogue. They take up positions and act as each other's counterpoint, without which no drama could take place. The drama is not a random exercise, its beginning lies in the planned open spaces and the impact this creates when inserted between avenues of apartments and commercial buildings.

The Landscape Architecture

Sophie Wolfrum

The urban design of Scharnhauser Park dwells upon the landscape motifs of the summit and the south facing slope. A few design rules ensure the distinctiveness of the five urban neighbourhoods whilst allowing an extent of freedom in their individual development. The design of the public open spaces forges the specificity of the urban location.

A cluster of tower blocks congregate at the summit and enjoy the prime views. The south facing slope with its distant views of the Swabian mountains and the Filder plains is articulated by the stepped terraces and green pastures between them. The arboretum at the summit and the formal avenue leading to the new tram station, combine with the long landscaped spine to provide the open spaces for public life.

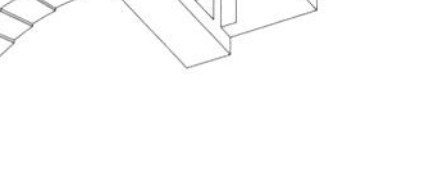

Landscape spine

Public buildings and residential apartment blocks line the long green landscaped spine on either side like a racecourse. The view from the summit opens out far across the fields to the east. The stepped terraces form the spine of the entire development enclosing the centre with a slight curvature until it reaches the open fields.

Architectural Discussion

Sophie Wolfrum

This region is famous for agriculture due to the fertile soil and an unusually shaped but delicious cabbage. There were also royal stud farms for breeding Arab horses and fields of linseed and lavender. In the 1930s the Luftwaffe bought the land for a commercial airport but after surveying the terrain they found it was on too much of a slope and moved the airport further south to where Stuttgart Airport is today and built a military airbase here instead.

The airbase became a US airbase after the Second World War, and when the Americans left in the 1990s there was an architectural competition held for master planning the new town of Scharnhauser Park in Ostfildern. The competition brief called for lots of green spaces and generous access to enable the town to grow into a city sometime in the future. There were to be new schools and a new town hall. Our plan was to have a very dense urban structure that is interlinked with the landscape, encouraging panoramic views onto the outlying mountains.

The offices and commercial buildings were located along the side of the main feeder road through the town which would screen off road traffic noise from the residential quarters behind it. There were to be no main roads through the residential areas. The residential areas were divided into subcommunities of different densities and housing types. The buildings were generally kept to four storeys, and many have duplex apartments which are popular in Germany. There are some tall tower block clusters in one area to the north which are more exclusive and command a higher price tag. Within a former dormitory zone for military personnel there is a mix of small and large apartment buildings. All the housing blocks are orientated southwards for maximum solar gain and enjoy a garden on the lower floors or views of the hills and landscaping on the upper floors. As the plots were very expensive the housing density had to be high to make them affordable. The introduction of courtyard gardens, landscaping and open spaces was arranged to liberate such dense massing.

Stepped terraces

Detail of drainage outlet

The landscape concept was to create a central terraced grassy slope that was 40m wide and 1km long running as a green spine through the middle of the conurbation. The drop was only 20m or a gradient of 2% over 1,000m which really does not warrant a staircase but we liked it for architectural reasons and for creating surface water drainage lagoons. The slope is hardly noticeable without the stepped terraces that break the green spaces at intervals along its length. We wanted to show the slope and not the flatness of the ground and this was one way of doing it. From the top of the slope looking down it is a green carpet cutting through the new town, from the bottom of the slope looking up it is a solid wall of stepped terracing. In the bleak winter when the ground is covered in snow and you cannot see much of the terraces it is still enjoyable to walk the kilometre.

All the rainwater from the nearby estate roads and apartment roofs is piped to flow into the hollows of the grassy slopes between the terraces to become temporary rainwater lagoons during prolonged rainfall. The grassy hollows fill from the highest level and any surcharge spills over the notch in the lower terrace steps to gently cascade into the next grassy hollow. The grassy hollows are soak-aways and will eventually drain the lagoons when the storm has passed and return to being grass covered slopes once again. This was a very pleasing and environmentally friendly way of disposing of the surface water. We had a water engineer who designed the drainage system. It is delightful to see children playing in the shallow lagoons after a storm, floating their boats or just paddling.

We choose white precast concrete for the terrace steps, the culverts and rainwater interceptors and sidewalks. It was inexpensive, durable and resembled a marble-like material. There was an option to use grey paving slabs but the council preferred white which was 10% more expensive. White is a better contrast with green. We referred to it as reconstructed stone to avoid the negative reception concrete often receives from the public. We also imported some large boulders of marble and placed them along the sidewalks near the terraces, and in some ways this reinforces the suggestion that the precast paving slabs are marble.

The central channel formed on the steps becomes an overflow and conduit when the lagoons are full but most times the steps are dry and the channel is just a surface feature. The choice of grass has been a major exercise in selection. Expert advice was sought from the turf and sports industry. There are three different grass seeds in the mixture to create a hard wearing, drought resistant covering that would be quick to recover.

Leading from the new tram station which has a curving concrete footbridge and sparkling bush hammered concrete balustrades, there is a paved arbour of paulownia trees with a children's play area, a network of drainage channels and rows of low level lighting. This formal piazza with flowering trees sits at the head of and looks down the long landscaped stairway. Along the sides of the green slope deciduous trees were planted which will grow to screen the hard edges of the apartment blocks that fringe its verdant pasture.

Precast terraces and central channel looking uphill

The illuminated hardstanding towards the tram station

Arboretum

PROJECT DATA

Client: Stadt Ostfildern
Architect: Janson + Wolfrum Architekten
Construction Supervisor: GP-Mack
Landscape Contractor: ARGE Link/Fischer
Water Engineer: Atelier Dreiseitl, Überlingen
Precast Manufacturer: Stangl AG
Completion: 2002

View downhill dominated by the green of the grass lagoons

Piazza and formal planting

SYNAGOGUE AND COMMUNITY CENTRE, DRESDEN

Wandel, Höfer, Lorch + Hirsch

Location

Dresden was once considered the Baroque capital of Germany and the Florence on the Elbe – until one fateful night in 1945 when the Allies bombed the city to rubble. It is slowly but surely being rebuilt and restored. The synagogue is located close to the historic centre of Dresden, on the south bank of the River Elbe. On arrival in the city centre head for the old town and the river. As you arrive at the river's edge you will see the panorama of the historic old town on the opposite bank. The new synagogue is an unmistakable silhouette to the east of the Bruehlsche Terrasse.

Architectural Statement

Dresden is characterised by two destructions, the burning down of Gottfried Semper´s synagogue in the "Reichskristallnacht" on the 9th of November 1938 and the bombing of the historical city on 13th and 14th February 1945. The destructions are historically linked, yet the architectural consequences could not be more different. On the one hand the city is reconstructing historical monuments like the Frauenkirche establishing a false continuity and a problematic pretension of architectural stability. On the other hand the new synagogue represents an attempt to investigate the conflict between the permanent and the temporary, between the old and the new order.

Inserted into the sloped topography of the site a central courtyard acts as a connecting element between the various uses of the synagogue and the community centre. Physical coherence is maintained by the use of a continuous material, precast concrete comprising cement, sand and aggregates. Each building however has a character of its own: The synagogue is a concentrated place of worship and meditation, its structure relating to the Elbe River and becoming part of the Dresden skyline. The community centre refers to the urban fabric and addresses the centre of the city.

Exploring the implications of stability and fragility, the architecture of the synagogue is characterised by a material dualism: a monolithic structure of precast concrete stones and an interior structure of soft metallic textile. The twisting stone structure of the synagogue follows the geometry of the site and the requirement of an orientation towards the east. The complex, curvilinear volume is based on a gradual shift of 41 orthogonal layers, formed by elements of 120 x 60 x 60cm of precast stone. By function and infrequent use this building has been given temporary building status and does not require to be insulated.

In contrast to the overall monolithic structure, the interior of the synagogue is framed by a smooth metallic textile. Suspended from a concrete ceiling grid it delineates the basic space of worship. The brass textile, developed with a clothing manufacturer, provides a specific aura of golden light. Wooden furniture characterises the interior of the "tent": a balcony, pews, the bima (a lectern) and the Torah shrine in the east. The position of the central religious elements evokes the spatial conflict of the synagogue: both a longitudinal and central space.

South elevation

Architectural Discussion

Nikolaus Hirsch

The synagogue refers in colour and texture to the sandstone construction of historic buildings in Dresden's centre, for example the catholic Cathedral, the Zwinger Pavilion, Semper's Opera House and many others. But it was not built of the same natural material. The sandstone of Dresden is a warm grey which has a high percentage of crystalline structure, turning the stone black over time. We developed an artificial stone, a reconstructed stone from engineered modern concrete that will always keep its durable warm grey colour and that is structurally more efficient.

The precast stone that we chose does respond to the historic buildings of the old city and also echoes the prefabricated concrete of the apartment blocks of the 1960s and 1970s close by. Precast concrete as artificial stone bridges both urban contexts. The finished colour of the concrete was arrived at after many trials and samples were made. It was not true that we only introduced reconstructed stone because natural sandstone was too expensive. Dresden stone was not suitable because it was brittle and much weaker than precast stone; it would have required the wall to be 1000mm thick instead of 600mm, thus becoming more costly. We wanted to use a contemporary material.

Community centre building

It is important to mention that there are two buildings on the site. It is a plot that is 26m wide by 110m long. It sits on a bluff on the banks of the Elbe. The concept of two separate buildings, one for worship and the other for community activities was a new interpretation of the competition brief. The brief required one building that integrated worship and social activities under one roof with a common entrance and foyer. As it was a long narrow site, we split these two functions and placed the synagogue at one end overlooking the river to be seen as a landmark. At the other end the community building links with the city fabric. Between these two buildings is a public space for both the local residents and the Jewish community to enjoy as they enter the café on the ground floor of the community centre.

The two buildings have quite different purposes. The community centre is used every day and is designed conventionally with a heating system, insulated walls and service installations. The structure is an in situ concrete frame with two suspended floors and a roof. It has a full-height glass curtain wall on the inward elevation facing the courtyard, the other three sides are wrapped externally with precast stone to exactly the same detailing and coursing as the synagogue, except that they are punctuated with window openings and the stone thickness is 120mm. The position of the windows may appear random, but in fact they are precisely located according to the logic of the interior spaces: high position for storage rooms and toilets and small and low as they follow the staircase up the building. The effect is that they do not dominate the solidity and mass of the façade. The community centre is a rectangular box 26m by 27m in plan and 15m high. The open space of the courtyard, the outline of the destroyed Synagogue that once stood on the site whose imprint is covered with shards of glass; the low boundary walls, the plinth and the entry steps are all made from the same precast concrete.

The synagogue is a very specific development. It is orthogonal at the base and fits squarely across the width of the site with a 26m by 24m plan. The walls then rotate incrementally as it rises until the axis of the temple faces east in holy alignment. Each stone course remains orthogonal yet turns layer by layer in a slowly rotating square until the volume has a 9° twist. From the base to the top of the external corner, one face cantilevers outwards 1.5m while the negative side cantilevers inwards 1.5m. Each layer rotates by 5cm at the corners, each precast stone is 1,200mm long, 600mm high and 600mm wide, with successive layers alternating as solid and then hollow blocks until all the layers are placed. The ribbed concrete roof slab ties the top of the walls together and reinforcement placed in the hollow block courses act as ring beams.

We were interested in developing a complex geometry that is based on a simple principle: only one form of stone from one mould for the whole building. The masonry was done by starting at one corner and setting the face of the next course back 50mm or forward 50mm from the stone below, the line of the new layer could be set out using a string line. This was a simple and effective way to set out the building but we only found out about it when the masons began the construction work. The blocks have recesses on the end faces so that they interlock when bonded with mortar. The depth of the horizontal joint was maintained by placing a piece of thin rope in from both faces and dabs of soft metal shims to support the block while the wet mortar hardened. The rope was removed when the mortar had hardened and the joints were then pointed. It was critical to keep the joint line true and the edges of the block in a line, as this would give a crisp shadow line to emphasise the stone layers.

Brass textile translucent curtain

Tented temple

Suspended textile curtain and ceiling roof light

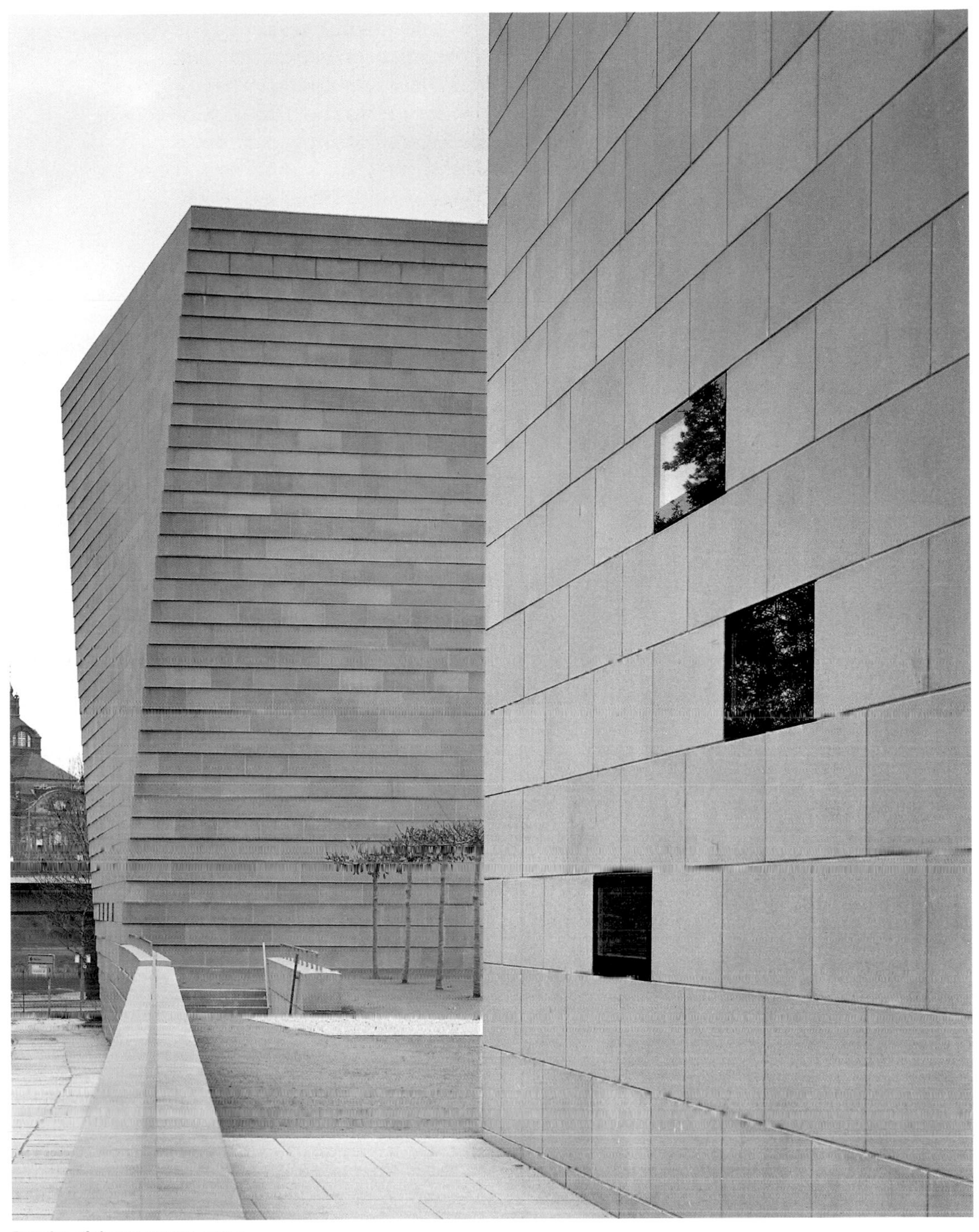

Rotation of the synagogue corner
along the south elevation

Draped fabric roof of the tent

Ribbed beam structure of the roof

One geometric problem that took some time to solve was the perforations that had to be made inside the blocks to integrate the services and rainwater pipes within the wall thickness. The hole was located in the block near the middle of the wall where there was the least rotation. Nevertheless because the coursing was stretcher bond the hole was in a different place for blocks in successive layers. Consequently the holes had different positions in the mould. The precision and accuracy in the manufacture of the concrete blocks, the symmetry of the alignment using the string line made the erection of the wall straightforward much to our surprise. Overall it took the contractor just 15 weeks to complete the building including the roof slab.

To arrive at the colour of the precast stone took a lot of experimenting. Unless you do this you will have no security of the surface texture, porosity and colour. The precast manufacturer was given a reference sample at tender stage for pricing. A number of sample stones were batched and once the colour had been settled using one of their standard crushed rock sands and a blended cement, the surface texture was finalised using acid-etching. Sand blasting, retardation and water jetting was tried but the finish was not as good and as consistent as acid-etching.

We were working to a tight building programme so this meant that the material supply was critical. It was agreed that as each batch of reconstructed stone was cast and hardened they would be sent to the site and installed layer by layer.

Stone courses of the synagogue

Laying the heavy precast stone blocks

Special stone blocks with location reference

Construction of synagogue building

We appreciated that the precast stones were not uniform in colour. It was a process that was sensitive to changes in air temperature and curing conditions at the time of casting. These subtle, random changes of tone added character to the building. We wanted each elevation to weather and the building to pattern and stain with age so that it experiences these responses to the environment and not remain a plastic synthetic colour.

In contrast to the monolithic structure, the interior of the synagogue is a tactile space. Suspended from a concrete ceiling grid it constitutes the basic space of worship. The brass textile, developed in collaboration with a clothing manufacturer, provides a specific radiant light. The transparent metallic screen which you can walk around refers to the tabernacle and the temporary structure described in the Torah. The seating, the gallery and the construction of the holy ark are made from oak, the worshippers sit in pews facing the holy ark enclosed by the tent. The metallic fabric is also draped across the concrete ribs of the roof slab to enclose the tent, while the lights are suspended by thin wires from the roof to illuminate the space of the temple.

The synagogue building has been classified as a temporary structure so it does not have to be insulated. It is only used on the Fridays and Saturdays during the Sabbath. The wall on the outside is the wall on the inside. We compared the cost-benefit of insulating the building and constructing an inner wall to that of a monolithic un-insulated structure with a heating system. Only after 50 years would we recover the money for the insulated scheme, so the client opted for the bare wall structure with a specifically developed heating system that you see today.

PROJECT DATA

Client: The Jewish Community of Dresden
Architect: Wandel, Höfer, Lorch + Hirsch
Structural Engineer: Schweitzer Ingenieure
Services Engineer: Zibell Willner und Partner
Main Contractor: Bau Nossen Stetzler
Precast Manufacturer: Decomo, Belgium
Completion: 2001
Construction Time: 2 years

Corner twist

Precision of joint line

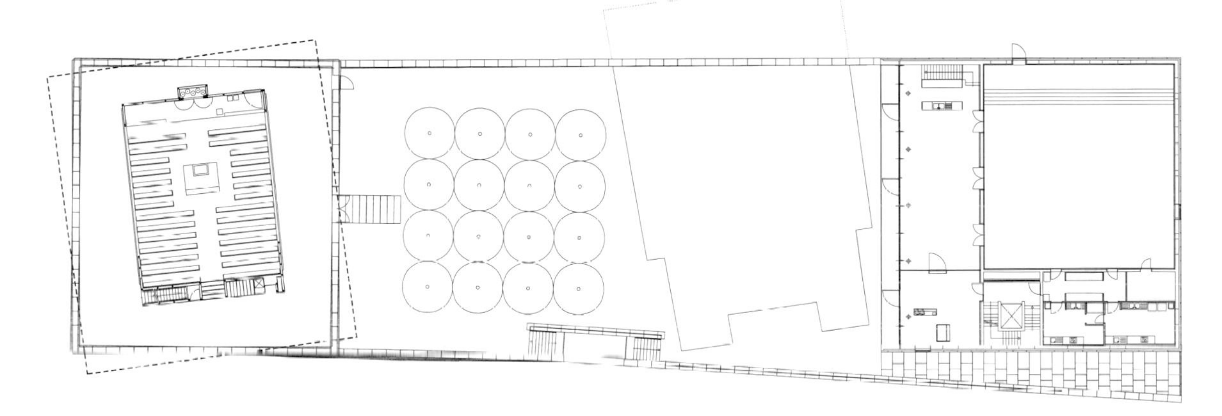

Site layout

5 10

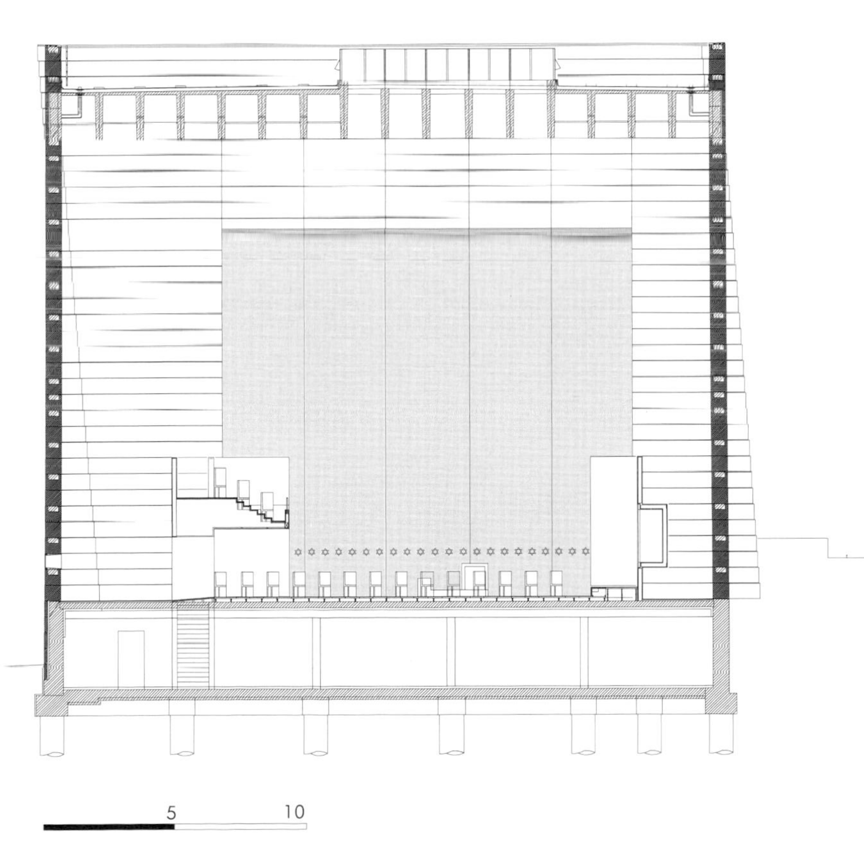

5 10

Section of synagogue building

Moulds for precast blocks

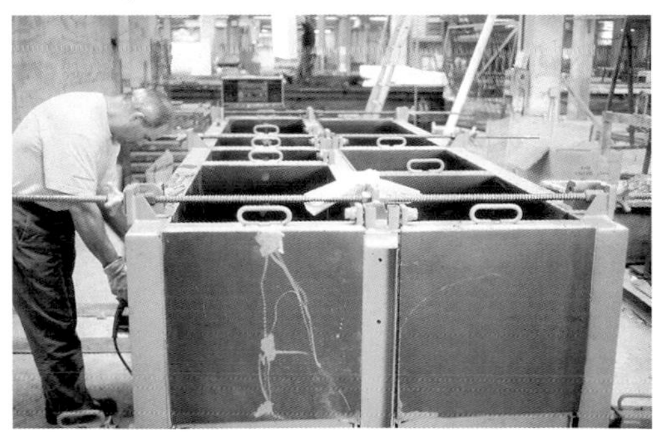

MEXICAN EMBASSY, BERLIN

Teodoro González de León and Francisco Serrano

Angled blade beam wall of the main elevation

Location

The embassy is located on Klingelhöferstrasse in the heart of Berlin. It is about a 20 minute taxi ride from Tegel Airport and a 40 minute taxi journey from Schönefeld Airport. You can also take Bus Line 100 from Zoologischer Garten Railway Station.

Architectural Statement

The new Embassy of Mexico is located on Klingelhöferstraße in the region of Tiergarten, a green belt space with a famous memorial that marks the historical axis of Berlin. The building occupies a plot of 1,300m^2 bordered by Rauchstraße to the north and adjacent to the celebrated curved green frontage of the Scandinavian consulates. Along the street frontage the white façade is dominated by a series of 18m high vertical blade beams that are angled in repose and curvilinear in formation, framed by a vast portal structure. Where the two planes of the blade walls overlap, the minor one tucks behind the major one to form the entrance vestibule. The gaps between the blade beams produce remarkable transparency, yet seen at an angle they create a solid visual barrier, developing a monumental dynamic.

The openness of expression, the symmetry and transparency of the lamina wall, the height and scale of the building symbolise the spirit of Mexican architecture. The entrance vestibule leads into a light-filled central atrium, a luminous white walled concrete drum 18m high and 14m in diameter which could be described as a vast chiminea with a glass topped roof or a concrete colander with portholes. A cascading planter terrace to one side of the atrium sits below glass link bridges that lead from the glass walled lift to the upper floors via doorways cut into the drum wall. The cylindrical atrium is the vertical axis and hub of the building, a conduit unifying the spaces within the building. The double-storey high reception area adjoining the atrium creates multi-functional spaces for receptions and cultural events. North of the atrium on the ground floor is the information centre, the public consultations rooms and the library. On the first floor there is the consulate department. The upper floors located along the north and front elevation of the building contain the administrative offices and the office of the Ambassador. There is a garden terrace on the roof with good views of the surrounding area. There are no plant rooms or mechanical boxes to obstruct the roof panorama, they are all housed in the basement which has an underground car park. The new building not only services all the consulate needs but accommodates conferences, exhibitions, cinema, theatre, music performances and diplomatic receptions.

The architects Teodoro González de León and Francisco Serrano won the design competition ahead of eight of Mexico's finest architects. The building had to represent Mexican values, the realities of a modern evolving society and to integrate itself into the urban environment of the Tiergarten district. Since its opening in November 2000 the building has received thousands of visitors and has been quickly added to the list of fine modern buildings to be seen on architectural tours of Berlin. It has had a positive effect on bilateral relations with Germany and has undoubtedly helped to improve commerce and trade between these two nations.

Architectural Discussion

Heinz-Dieter Witte & Pirkko Helena Petrovic, Assmann Beraten und Planen

In the mid 1990s González de León then in his seventies and Francisco Serrano in his sixties teamed up to enter the Mexican Embassy design competition. González de León, who had worked with Le Corbusier in Paris in his formative years, upon his return to Mexico in the 1950s soon established his reputation for the design of large public and residential buildings. Francisco Serrano who had started in his father's practice was responsible for extending and developing the campus of Mexico University and had built many hotels and office buildings in Mexico over the years.

The planning regulations restricted the height of the building to 18m, so that it would conform to the scale of the adjacent Scandinavian Embassy complex. The site was bounded on two sides by streets. The concept was to create a dynamic façade made out of thin blade elements that curved in two planes; the material was bush-hammered white concrete. The interior space was to be open and vast with a style that was typically Mexican. The focal point is the four-storey high atrium designed as a white-walled cylinder of concrete which is perforated on the south-western face by portholes of plexi-glass. The portholes and the glass-canopied roof maximise the sunlight into the atrium and the terrace of planters for ferns and evergreens on the ground floor. The atrium connects the lift to all the upper floors and at ground level it leads onto the double-storey high exhibition hall and reception areas that can be interlinked to create a larger function area. Throughout the building the exposed walls and beam surfaces are made of white bush-hammered concrete. For this material resembles a natural stone of Mexico, its surface improves light reflection and keeps the building cool in summer.

Design sketch

Entrance

The vertical blades were precast but the portal frame and the atrium cylinder were cast in situ to exactly the same finish as the precast concrete. The quality of the in situ finish was so remarkable that it was difficult to distinguish from the precast panels. A sample of white bush hammered precast concrete and the mix design details were given to each of the tenderers with the tender documents. As the representative of the Mexican architects we made the planning application and carried out the contract administration duties. As part of our induction we made two trips to Mexico for workshop, design briefings and to visit the architects building to get a feel for what they required. The surface tooling of concrete in Mexico is all done manually by stone masons. To replicate the same effect in Berlin it was the original plan to bring over Mexican stonemasons and have them carry out all the bush hammering but it was not a success. The precast manufacturer chose to employ local labour and use pneumatic impact hammers to create the exposed aggregate finish the architect required.

The series of façade blade beams to the front and side elevation were precast north of Berlin and transported two at a time and erected in place. The front blades were tilted to a prescribed angle and connected at intervals to the edge of floor slabs and tied to an enormous in situ capping beam over the top of the structure. Each blade beam had a unique profile and length although the thickness and width were the same. They were angled differently against the building and located in a gentle curve. The blade beams on the side elevation were of uniform height and were in vertical alignment.

The portholes cast into the in situ drum wall were made of plexi-glass, the inside vacuum was kept moist-free by the inclusion of silica gel crystals. The ceilings were covered with gypsum plaster board, the offices were carpeted with a special fabric, the ground floor was laid with yellow/grey marble and the façades received anti-graffiti coating close to ground level. The services are contained behind a false ceiling, lighting is on the wall and not overhead, floor screed was laid over the under floor heating pipes. The outstanding aspect of the exposed concrete was the edge details of the drum wall panels.

Form top to bottom:
Forming the drum of the atrium
Concrete drum wall and portholes
Bush hammering
Erecting blade beams
Blade beam connections

Precast Construction
Christian Rymarczyk, Geithner Bau

The factory is located in Groß-Ziethen some 70km north-east of Berlin. All the precast elements were made here, including the 18m long blade beams. The concrete mix was based on what the ready mixed supplier produced for casting the in situ drum walls. We had to keep to the recipe to ensure a close colour match with the rest of the building, which was cast in situ.

The white cement was supplied by Dyckerhoff cement, the sand fines and coarse aggregates consist of a crushed white marble from the Lagerfeld quarries. The coarse aggregate size ranged from 8-35mm, the fines from 0-2mm. The distribution of particle sizes throughout the sand and coarse material was kept constant for all the precast work. The cement content and water/cement ratio were fixed for every cubic metre of concrete that was batched, and plasticiser was added to ensure uniformity of flow.

The materials are weigh-batched in one cubic metre lots. The water is controlled to within 1 litre of specification and aggregates to about 10%. Grading curves and sieve analysis are carried out regularly to monitor material quality and wet density and water content tests are conducted on the fresh concrete to check the actual water content in the mix. The precast elements are inspected when they are removed from the moulds, and if there are too many surface imperfections for example due to blow holes, they are discarded. On the Mexican Embassy contract not a single precast beam was rejected. It helps that the surface was bush hammered as this removed a lot of minor surface irregularities.

The concrete was mixed in a horizontal turbine mixer in a purpose-built batching plant. Concrete was discharged into a skips that moved on rails to the covered formwork yard. The blade beams were 250mm thick and 900mm wide and were made in wood lined 18m long beam moulds. It required four skip loads to fill each mould which took 30 minutes. Internal poker vibrators and clamp-on external vibrators compacted the concrete. For heights of up to 1.4m conventional concrete mixes are used but for larger panels which are free-standing self-compacting concrete is specified. Most of our cladding panels are cast on the flat. Self-compacting concrete is two to three times more expensive. It is not the cost of the specially graded sands and filler that we buy in and have to carefully monitor, it is the amount of testing work that makes it so expensive. For conventional

Portal framed blade beam elevations

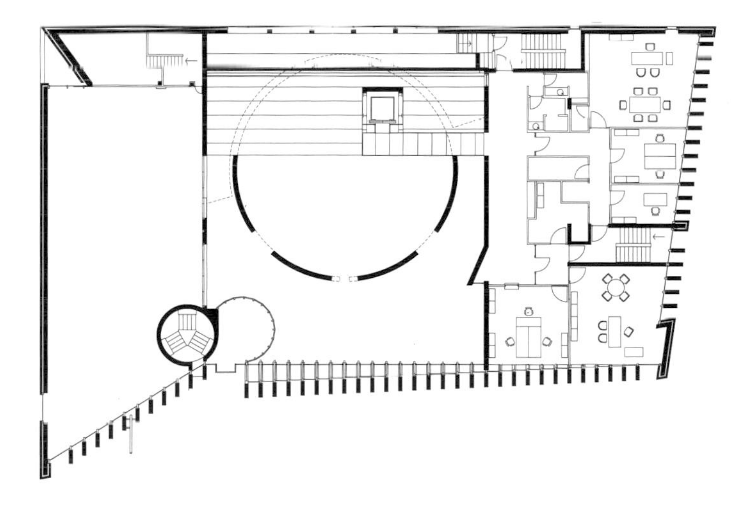
Ground floor plan

Section through atrium space

concrete mixes we deploy a technician for 30 minutes per mix per day, on self compacting it is 2 to 3 hours/day/mix. We are currently casting over 2,500 monolithic blocks of black self compacting concrete e for a memorial garden in Berlin dedicated to the holocaust and designed by the architect Eisenman. Each piece weighs between 5 and 10 tonnes and stands up to 5m high.

When the beams are removed from the forms they are inspected and taken to the treatment area for bush hammering. The work starts once the concrete is 7 days old. First a shallow saw cut is made a set distance from the edge or corner of the panel to create a margin. Two teams of operatives both start bush hammering from the middle of the beam and work their way to each end. It takes about 15 hours to bush hammer and finish a beam, each operative achieving 5m^2/day. We employed 15 men to carry out the bush hammering work in the stockyard.

Two 18m beams are taken to site by lorry and erected into position using two mobile cranes. One crane lifts the beam off the lorry in the horizontal position using nylon ropes, the other picks up the beam from the top end via the protruding reinforcement bars and positions it on the foundation plinth at the correct angle. It is held in place until the bolts are secured and it is connected to the edge of the floor slab. When the major wall of blade beams were in position the formwork for the long, wide capping beam was assembled, then cast with white in situ concrete and bush hammered on site. The minor blade wall which sits below the capping beam connects to the underside of the concrete roof slab.

Entrance lobby

White walled atrium cylinder and portholes

PROJECT DATA

Client: United States of Mexico
Architect: Teodoro González de León and Francisco Serrano
Architect's Representative in Berlin: Assmann Beraten und Planen
Structural Engineer: Assmann Beraten und Planen
Services Engineer: Assmann Beraten und Planen
Main Contractor: Groth Gruppe GmbH
Concrete Subcontractor: Hochtief AG
Precast Manufacturer: Geithner Bau
Completion: 2000
Construction Time: 11 months

Glass canopy over atrium

'Blade wall' elevation on the side street

Main elevation

10 CROWN PLACE, LONDON
MacCormac Jamieson Prichard

Street elevation

Location

This office building is at the corner of Crown Place – a pedestrianised street that was formerly Clifton Street – and Earl street, next to the Broadgate Development in the heart of the City of London. It is a short walk from Liverpool Street Station.

Architectural Statement

The building has been designed as a modern reinterpretation of the 19th century warehouses which are typical for the surrounding conservation area. The arrangement of paired columns and beams in fine precast concrete, of a quality resembling Portland stone, together with slender intermediate stainless steel columns, follows the elevational composition of many of the building's older neighbours which are built in brick, stone and cast iron. The façade is strongly modulated by projecting glass bays which sit within the precast concrete frame. The ground floor glazing is set back to give greater privacy to the occupants whilst allowing natural light to drop down into the basement along the perimeter.

The lifts, stairs and service cores have been used as buffers against the party walls of the adjoining properties. The stair towers are clad with a slate rain screen punctuated by a thin vertical glass lantern running the full height of the tower. The main entrance is marked by a glass canopy which cantilevers out over the pavement, designed by the artist Dan Chadwick. A large projecting precast concrete cornice at fourth floor level completes the composition, shielding the two-storey high glazed mansard from street level. The interior of the building is composed in an elegant palette of materials using French limestone, Brandy Crag slate, white plaster, Venetian render and pale oak.

It was important to demonstrate to the planners and conservation officers that the design of the proposed building would integrate successfully into the conservation area in Shoreditch. The scale of the building was appropriate to the adjoining buildings along both Clifton Street and Earl Street. The design was to be of high quality, capable of setting the standard for new developments in the area.

As a speculative office building, the 9m structural grid and 1.5m planning grid ensures flexible and efficient space planning for a wide variety of tenants.

Typical floor plan

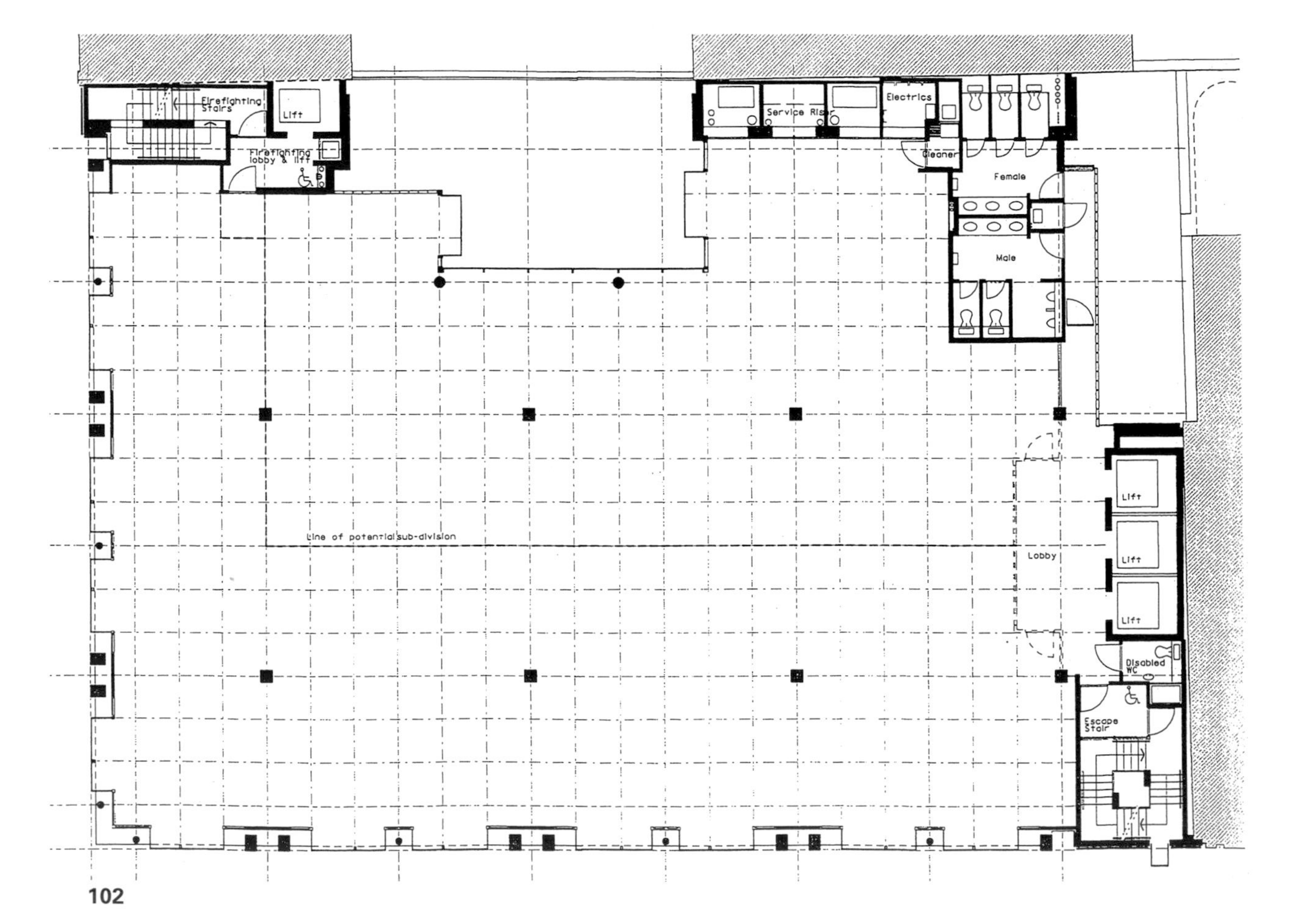

Building corner and cornice

Left: End elevation
on East Street
Right: Window box proud
of precast facade

Schematic drawing of precast edge beam and gutter

Main elevation

Architectural Discussion

Jeremy Estop, MacCormac Jamieson Prichard

A high quality contextual design was required to achieve planning permission and conservation area consent to demolish an existing building and re-develop the site. We made a detailed study of 19th century warehouse buildings, which are prevalent in the surrounding area, and designed the façade as an arrangement of primary, secondary and tertiary elements typical of these older buildings. The modern façade was built using precast concrete paired columns and beams punctuated by large bay windows which provide a well-lit interior.

The precast façade is not a simple cladding system but an active load-bearing frame that supports the perimeter of the floor slabs. During construction, the precast columns and edge beams were propped in position while the in situ concrete floor slabs were cast, tying them into the rest of the frame. Starter bars on the back of the edge beam units and UC sections projecting from the back of the paired columns were linked into the reinforcement bars in the in situ floor slabs. The columns were cast in storey-height lengths with hollow sleeves through the centre. They were joined on site by dropping reinforcement bars into the sleeved holes and filling with grout. Channels cast in the tops of the beams act as rainwater gutters, discharging the water through conical holes onto the roof of the bay window below (the holes are conical to prevent blockages by leaves and debris). Water then runs down specially designed channels located behind

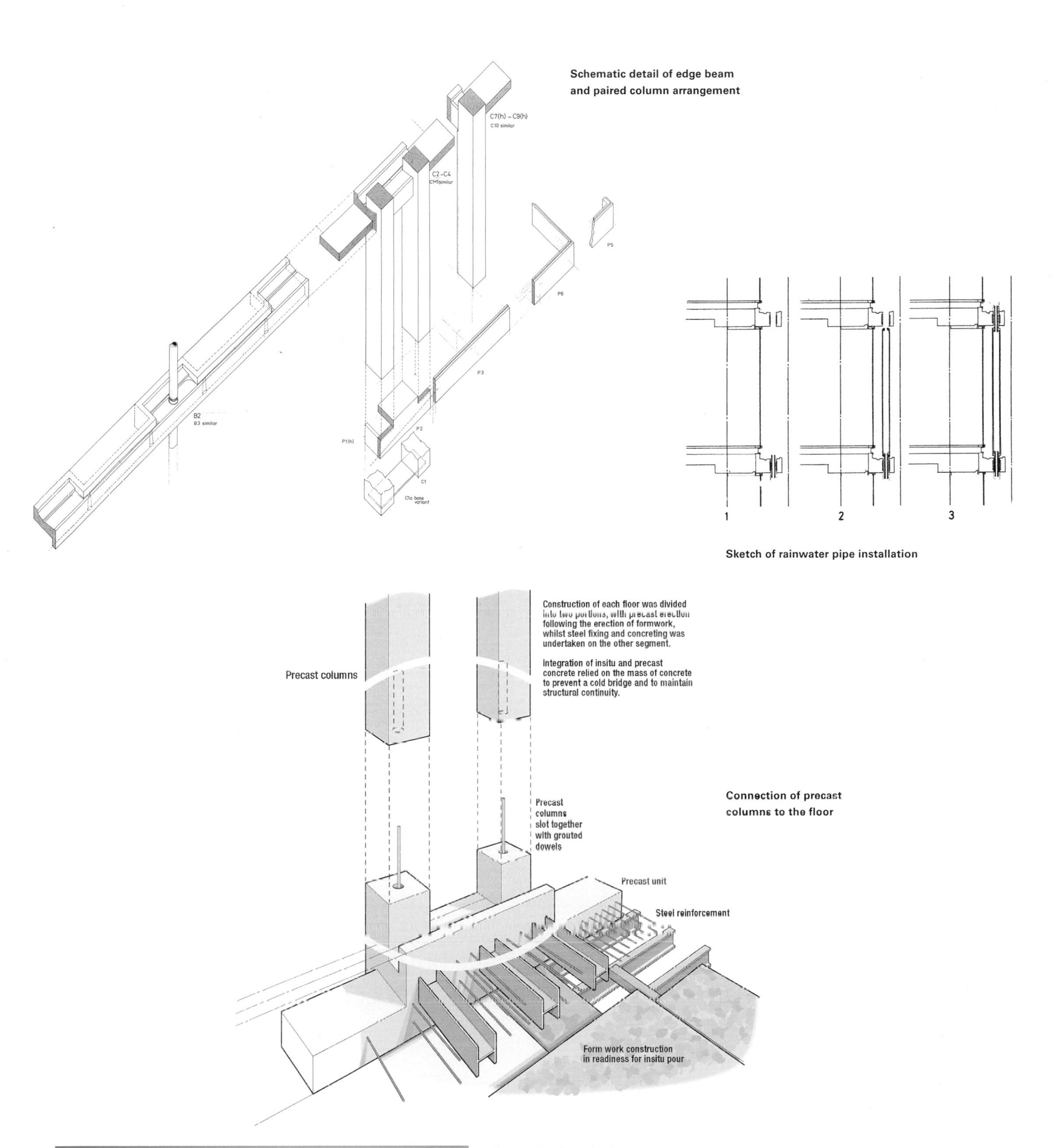

Schematic detail of edge beam and paired column arrangement

Sketch of rainwater pipe installation

Connection of precast columns to the floor

Precast façade set back from windows

the edges of the glass. The glazing is forward of the precast beams to minimise rainwater splashes onto the concrete which would stain it. The projecting cornice also provides good protection from the weather and explains why the precast surface still looks clean and unblemished after seven years.

The precast façade terminates at fourth floor level with a large projecting cornice masking the two-storey high glass mansard roof above it. The office floor plate was on a 9m square grid and provides very efficient and flexible floor space.

We have tended to work with Histons over the years because of the quality of their products. They used to be the only precast manufacturer to offer polished concrete finishes. It was Gordon Whitwell who was the driving force behind the company in the formative years, producing beautiful precast panels in the unlikeliest of surroundings – a converted pig farm near Wisbech.

We discussed finishes, profiles and panel sizes with Histons during the design stage and they influenced the development of the details on many of our buildings. We chose a white concrete made from white cement, crushed limestone aggregates with an acid-etched finish to all surfaces except to the tops of the beams with the rainwater channel which were polished to a hard wearing terrazzo finish. Together with the engineer Whitby Bird, we developed the load-bearing façade details on Crown Place. During the construction period we made visits to Histons factory in Littleport near Wisbech to inspect the quality of units before they were despatched, including a mock-up. Histons are a highly skilled company and we have great confidence in the quality of their work and their ability to meet our requirements. If there was any damage or scuff mark to panels on site a specialist team of repairers was employed to restore the surface to exactly match the concrete finish.

Mansard office space

Modelling of façade to echo 19th century warehouse vernacular: left original, right re-interpretation

Night view

PROJECT DATA

Client: Earl Estates Ltd
Architect: MacCormac Jamieson Prichard
Structural Engineer: Whitby Bird
Services Consultant: Ferguson Casley Rudland
Main Contractor: Willmott Dixon Construction
Precast Manufacturer: Histon Concrete Products
Completion: 1998
Contract Duration: 18 months
Total Cost: 8.3 million GBP
Gross Floor Area: 5,400m^2

Entrance lobby

Lift lobby

Glass lantern and slate-clad elevation of lift well

35 HOMER ROAD, SOLIHULL

Foggo Associates

Location

The site on Homer Road is an easy walk from Solihull train station which can be reached by trains from Birmingham New Street and London Marylebone. Solihull is served by good motorway links from the south via the M5 and M40, from the east and London via the M1/M6 and from the north via the M6. It is a 20 minute car journey from Birmingham City Airport and The National Exhibition Centre.

Architectural Statement

Homer Road was designed as a speculative office development providing 6,000m^2 of usable floor space spread over three storeys. The brief was for a high quality headquarters office building, maximising the openness and flexibility of the internal space and a low energy strategy.

Flexibility of the configuration and layout of the internal accommodation was a major design consideration. The building must be capable of being fitted out for both open plan or cellular space, offering secure and separate zoned areas within a floor using demountable partitions and provide security between floors for different departments or groups of companies.

Architecturally the aim was to create an elegant, transparent pavilion building with the ground floor floating above the natural slope of the site and the car parking. The building is surrounded by mature and semi-mature screen planting to provide a visual break from Homer Road, the road to the south of the site, and a footpath running adjacent to the long north-west elevation that leads to Tudor Grange Park. The main entrance is located on the shorter north-east elevation at the corner of Homer Road and the footpath, and has been opened up to become the focus of the new building and a gateway to the park.

The building is arranged on three rectangular floors each one 60m by 37.5m in plan with the building positioned closer to the junction of Homer Road and the footpath next to the site. The frontage reinforces the informal building line of the street. The floors are arranged in two 15m plates separated by cores and an atrium space. The atrium allows daylight into the heart of the building and is animated by glass lifts located at one end. There are two service cores at each end of the building, and they contain the fire escape stairs, toilets and service risers.

As part of the low energy strategy, an exposed concrete soffit was designed for thermal mass damping which also avoids the cost of a false ceiling. As the office space required to be used on a 24 hour basis it was necessary to introduce chilled beam ceilings and a displacement air system to optimise internal comfort conditions and integrate this into the exposed concrete soffit. The internal space exploits the precision and sculptural qualities of precast concrete construction.

The design of the external elements has been carefully developed using naturally finished, high quality materials. The elevations of the building are neutral, the transparency of the glazed elevation reflects the natural parkland setting. The atrium and the restaurant on the ground floor provide informal meeting spaces and are used extensively for group discussions by National Grid Transco who have occupied the building since it was opened.

Architectural Discussion

Steve Baker, Architect

The building was initially designed as a speculative development but very soon it became apparent that it was to be occupied by a group within the client organisation. Effectively the fitting out was purpose-built but the base building had to be designed to be flexible and adaptable throughout its useful life. The tenant was involved in the project from day one, so that we had to meet the client's brief as well as tenant requirements. The client was SecondSite Property Holdings Ltd, the property arm of National Grid Transco, the major gas supplier in the UK. They build on their own surplus land and sell the developments, usually to investment companies and pension funds.

Our first impression of the site was not inspiring. The plot, which had previously accommodated three Edwardian villas, is on a road leading into the town centre that was fronted with a hotchpotch of utilitarian, predominantly brick clad, residential buildings, hotels, a magistrates court and a police station with little architectural merit amongst them. Our immediate concern was that the local planning authority may not have been receptive to change or new ideas. Those fears were soon dispelled at our initial meeting with the senior planning officer who was a very enthusiastic supporter of the building and our design strategy from the outset.

Homer Road elevation

South-west long elevation and undercroft

Ramped main entrance

The building layout, the internal organisation and the external elevations were based on a 1.5m module. Having been given the target of 6,000m^2 of floor space, we studied a large number of options for floor plates, number of floors, large plan, small plan, low rise and high rise, and finally decided in favour of three floors on a rectangular footprint.

For an effective low energy building, a precast concrete structure with night time ventilation to cool the exposed structure is a logical solution. With a tenant requiring 24 hour occupancy of a building, night time cooling of the structure is no longer viable and mechanical cooling had to be installed. We developed the precast concrete structure to accommodate an integrated multi-service ceiling tile incorporating artificial lighting, acoustic panels, chilled beams and ceilings in lieu of a traditional false ceiling. The refinement of the precast elements, the modelling of the structural shapes and the high degree of repetition of units that was achieved was due to our close working relationship with the precast manufacturer.

The finish of the precast units was a key issue. We would have preferred the concrete to be unpainted but that may have caused a problem with uniformity of colour which can prove difficult during manufacture. This could lead to a higher number of rejects and further losses against chipped and damaged units, which would be reflected in a higher unit price. The painted solution was ultimately the optimum and most cost-effective one because surface colour is not critical and any minor repairs could be masked by the paint – it was also preferred by the client and tenant.

An office building in such a location outside a major city would not normally be capable of sustaining the cost of a high quality glass curtain wall system. We were able to make it fit the budget by keeping the details simple, modular and highly repetitive. There is fritting at high level for solar shading and there are internal blinds that can tilt and turn to control directional sunlight. By not having a false ceiling,

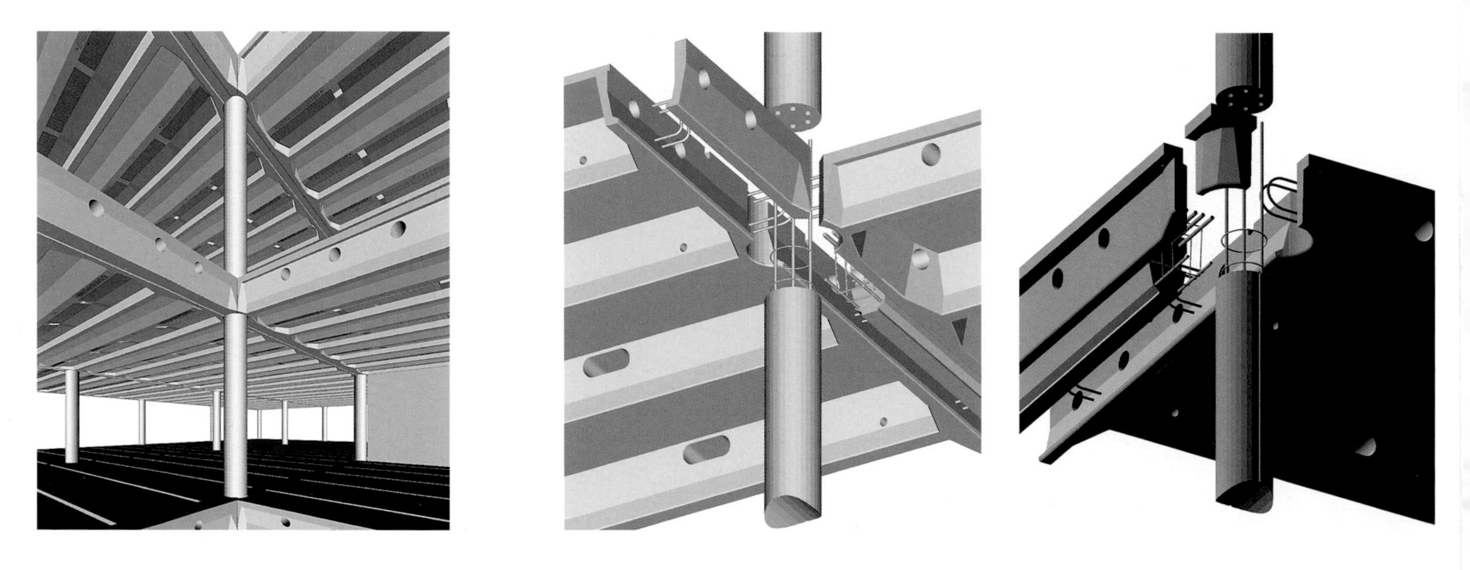

From left to right: Isometric of atrium corner and floor Internal and external corner connection

running the services within the precast coffers and passing circulating air through purpose-made holes in the edge beams of the atrium and building perimeter, we increased the daylight entering the building. The transparent glass maximises the views out of the building and reveals the edge beams and precast columns. We worked very hard to detail the edge beams to express the structure along the perimeter of the building.

Precast and Structural Considerations
Kimbell Grady, Structural Engineer

Because of the building's 7.5m column grid, one of our aspirations was to significantly reduce the amount of in situ concrete work that would be exposed, as such spans did not give the best surface finish. We set out to design the floors so that the entire soffit would be covered with precast elements including the edge and corner conditions around the building perimeter and the atrium. At first this design generated a large number of moulds which would be uneconomic. The modular floor and the repetition of the profiled soffit using double T and single T beams spanning onto a wider-shaped primary beam, were the results of further design development. We refined the rib and beam shapes until we were down to just seven types of moulds. The only in situ concrete visible was a 150m wide strip on the line of the primary beams, between the two halves of the precast beam shells.

The single Ts could be cast in the double T master moulds by blocking off one end; other units could also be adapted by adjusting other master moulds. The endplates of the double Ts and single Ts form one half of the primary beam shell that they span onto. The primary beams are then formed with in situ concrete after rebar has been positioned to complete the structural floor. It is a neat structural solution and very efficient with primary and secondary beams having an overall depth of 550mm with the slab 100mm thin and the secondary ribs running at 1.5m centres. We prepared a series of isometric drawings showing how the precast floor elements and in situ columns come together and issued them to SCC, the precast manufacturer. By introducing in situ concrete into the primary beams and columns we designed a sway frame to eliminate shear walls. In this way the contractor had complete freedom to start precast

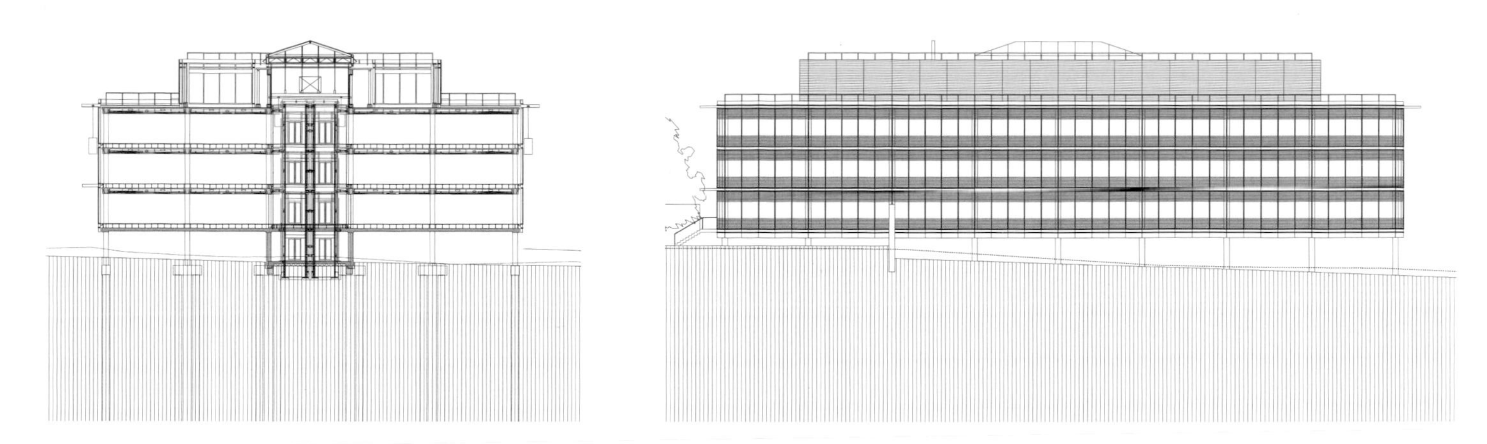

Typical section

North-west elevation

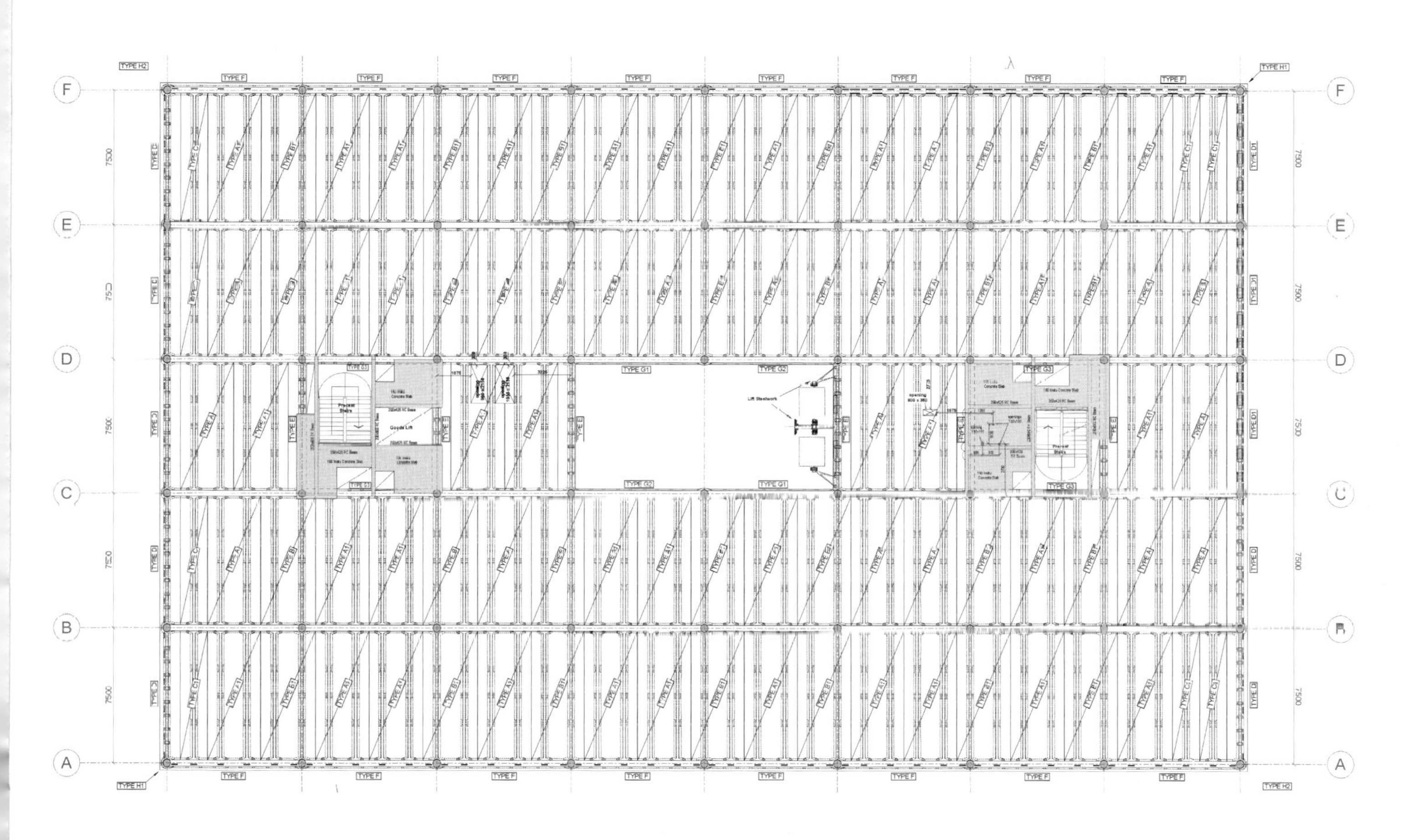

Precast double T and single T floor layout

Erection of fascia panels

Precast beam mould

erection on any part of the structure. A braced structure with shear walls would have required temporary bracing during erection and a more restricted sequence of work.

The precast manufacturer pleasantly surprised us by electing to precast the columns in single-storey lengths and connect them using grouting tubes cast into the columns. The starter bars of the lower column sleeve into ducts cast in the base of the upper column and the void is filled using the grout tubes. SCC had used this technique successfully on a number of precast car parks they have built and found it to be quicker. The grout holes on the column were concealed within the depth of the raised floor.

MIX CONSTITUENTS

Blue Circle Hope portland cement: 360kg
Tarmac Tunstead 20mm limestone: 722kg
10mm limestone: 481kg
Cheshire Sand Fourways Quarry: 352kg
Tarmac Tunstead limestone sand: 352kg
Water/cement ratio: 0.47
Target slump: 75mm

There is no basement to the building but there is an undercroft below the raised ground floor which was for a car park. The suspended ground floor was a cast in situ flat slab, so that while the pile caps to the CFA piled foundations and the ground floor were being constructed, parallel working was achieved with the precast columns and first floor double T being fabricated off-site.

There was no topping concrete to the precast units to tie them together, so the longitudinal edges of the double T had a groove which formed a shallow channel with an adjacent unit. Along the edge were cast-in steel loops at regular intervals, and a single rebar was then threaded through the overlapping loops and the channel grouted up. It is a system that has been used before and it works very well. There is no differential movement between precast units under load and we can develop diaphragm action using the in situ stitches in the primary beams, columns and between the rib units. Being precast it was precisely made and when they were in position on the staging they gave a flush soffit line. It was easy to see if a unit was slightly out of true level or line, and adjustment could be made on the propping. The tight detailing and refinement of the system paid off. The most critical area for tolerances was the primary beam and column connection. We did not want the column to be pushed out of verticality due to cumulative tolerances of the precast units. We ensured that there was enough play around each column to cater for this. If you look carefully you may notice that the columns and preformed holes they had to fit in the primary beams are not always concentric. There was enough of a shadow gap for that not be noticeable unless you stood directly below it and looked up.

SCC made the moulds out of ply rather than steel as we wanted sharp edges, neat fold line in the splays and good definition on the chamfered corners. The steel shutters tend to give a rounded edge and corners because of the nature of the fabrication. The plywood moulds were placed on vibrating tables to vibrate the fresh grey concrete mix. A few units had minor surface defects, blow holes were repaired and filled and masked by the white concrete paint that was applied on site later on.

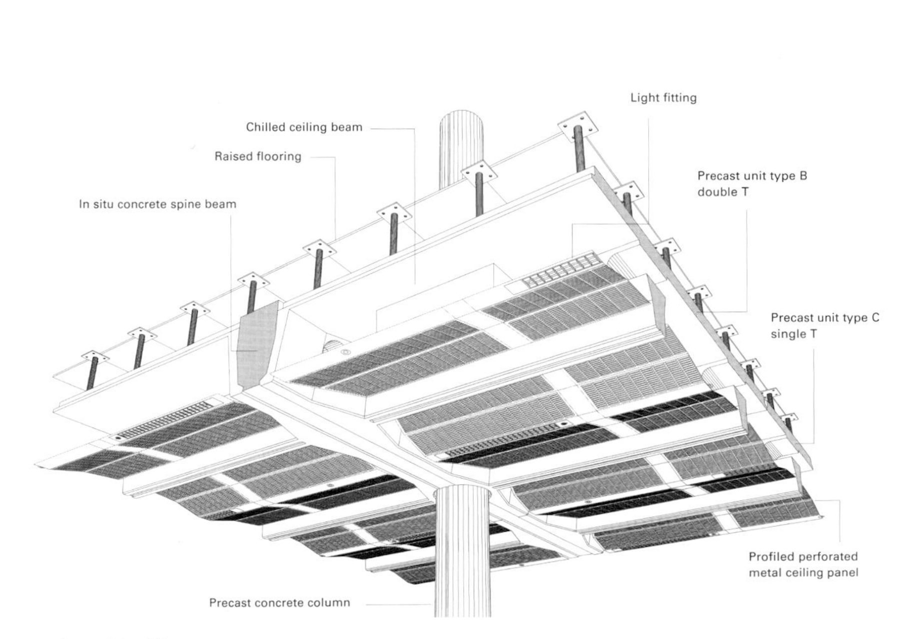

Isometric of floor and ceiling details

Precast structure exposed behind glass curtain wall

PROJECT DATA

Client: SecondSite Property Holdings
Architects/Engineers/Cost Consultants: Foggo Associates
Construction Manager: Bovis Lend Lease
Precast Manufacturer: SCC Ltd
Landscape Consultants: Hyland Edgar Driver
Completion: 2003
Construction Time: 16 months
Contract Value (including fitting out): 11,784,000 GBP
Gross Internal Area: 7,843m²

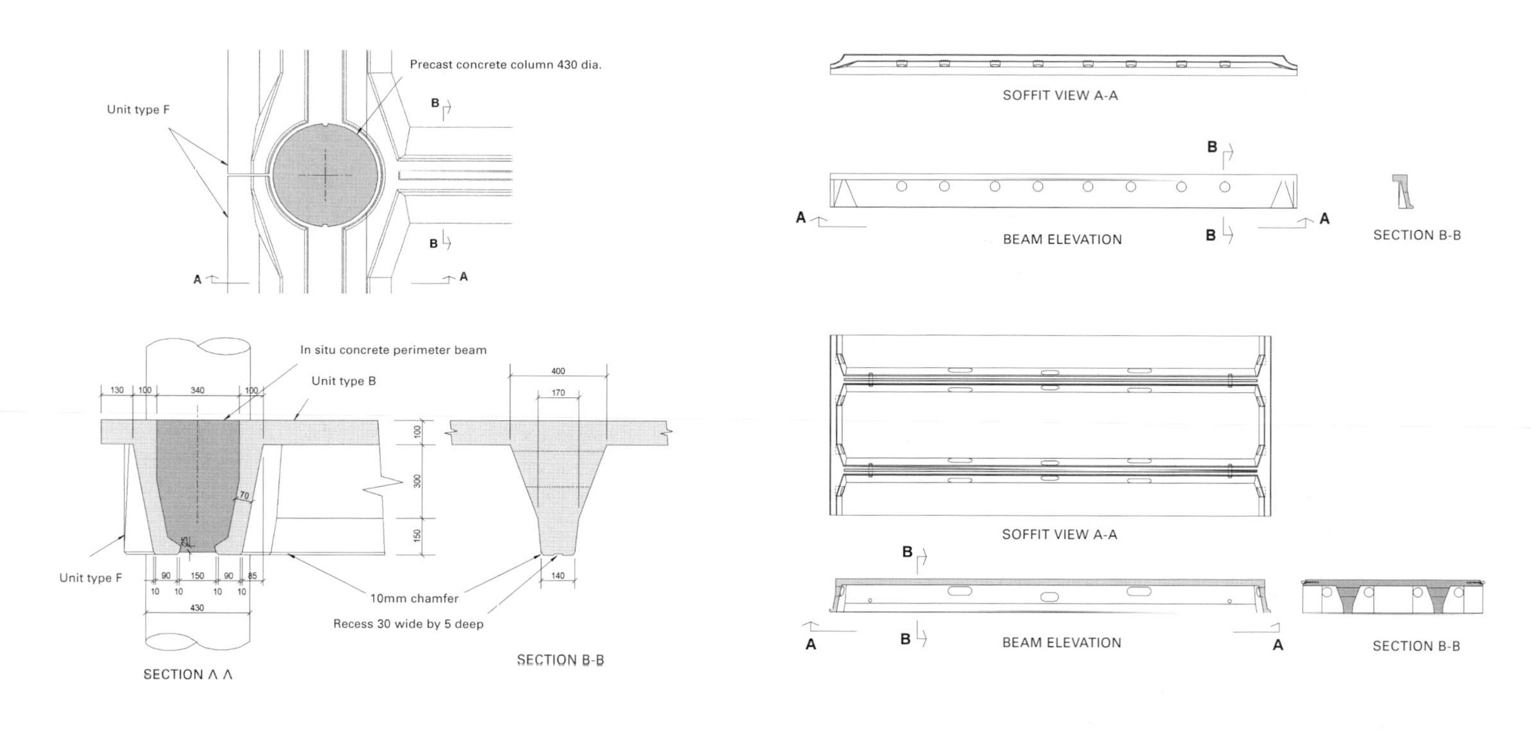

Details of in situ/precast stitch

Details of precast double I panels

Exposed precast beam soffit at atrium corner and slotted ceiling panel

Stacked precast floor panels

EXPERIAN DATA CENTRE, NOTTINGHAM

Sheppard Robson

Location

The site lies on the south-west corner of Ruddington Fields Business Park on the outskirts of Ruddington, 5 miles to the south of Nottingham, along the A60. It is a 20 minute taxi journey from Nottingham station.

Architectural Statement

The aim of the architects was to design a building that was secure, resilient and low rise and to provide a stimulating environment for both customer and workforce. The architecture expresses the status of Experian in the Global Information Solutions market and their ambition for the continued growth of the business in the coming decades.

The key to this was to design a substantial shield wall to prevent a forced entry into the heart of the building to access or damage the electronic information stored in special enclosures within it. The main elevation of the building divides the landscape in front of the building from the controlled secure zones within it. The cladding for the main external elevations and the enclosure of secure units within are the dominant features of the building. Their mass and form has been very specifically conceived and designed.

The front elevation which has been inspired by the aesthetic of stacked natural dry stone walling, is constructed of narrow precast concrete panels perforated by glazed slots that allow natural light into it. These openings allow views out of the building whilst controlling views into it. The substantial character of the front elevation is intended to express the security and prestige of the facility. The thermal mass of the façade will assist in stabilising the internal temperature.

Behind the main elevation is the main 'street' zone of the internal accommodation which contains the reception and entrance lobby, the media exchange offices, the café and break-out spaces. This zone is separated by a glass screen wall from the four large windowless chambers which dominate the interior of the building and which contain the Command Centre and the three Network Rooms. Between the 'street' and the secure internal chamber walls there is a 150m long main corridor that has a glass covered roof and connects with the other corridors. A lush reed-bed in front of the main elevation provides a contrast with the solid linear frontage and when it is fully established will create a sustainable wetland habitat for wildlife.

The external enclosures of the Command Centre and the Network Rooms are both structurally and visually robust. The Command Centre is clad with 3m wide by 7m high curved and patterned precast concrete panels. The fake joint lines on the panels hide the actual joints between them creating a monolithic appearance.

The Network Rooms are enclosed by 9m high by 2m wide precast units. The panels form the internal and external finish and act as bracing for the structure. The office spaces to the south and west wings of the building are enclosed with full-height curtain walling and solar-shaded with internal roller blinds and an external brise soleil.

Precast concrete has been used extensively in this project. Its properties guarantee modularity and speed of construction whilst allowing subtle play with texture, colour and shape. The precast application range in use from a rain screen to fully insulated walls and to panels with an outer and inner surface finish.

Our clients wanted to re-brand themselves through the architecture and style of this building which is unlike anything they have ever had built before. The new building has phenomenally improved the identity, saleability and success of the company since it opened.

Architectural Discussion

Tim Evans

The building is a modern fortress adjacent to open countryside, and has few people occupying the vast spaces inside it. Buildings like this in the USA are typically in remote locations and built like large featureless sheds. This site was close to the urban fringes of a city so it had to have a strong sense of design.

The client was concerned about how to protect the sensitive data inside and ensure that there was never an electrical or services failure that could literally melt down the hugely expensive banks of data and retrieval machines. No expense was spared on achieving these high levels of resilience, there were even back up systems to the standby equipment.

Our aim was to make the necessary security as overt as possible hence the design of a building with a solid impenetrable exterior.

The front wall to the building is 170m long, in front of it is a moat – a reed bed that filters waste water – and there is a security fence that runs along the site perimeter. We did not need a barrier wall to the rear elevation, which looks out onto fields, as it has no road access. The landscape architect designed a series of deep ditch-style 'ha ha' defences to the rear.

The architectural concept of the front wall was based on the regional tradition of dry stone walling. They are built with thin lengths of stone that are

Erection of 17 tonne curved precast panel for Command Centre

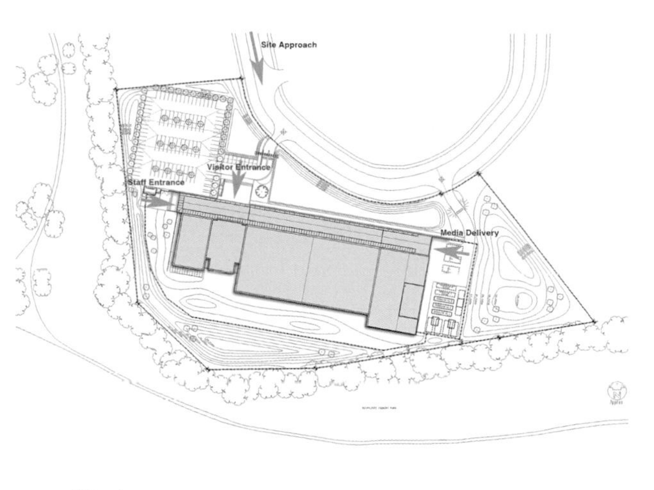

Site plan

Strong wall and slotted windows

Elevation with moat

chipped and shaped to wedge into the dry wall. We also likened the wall and the slotted windows to the old computer punch card and punched data tape once used for storing information. The 7.4m high exterior wall is composed of narrow stack-bonded precast concrete panels 465mm high and either 6m or 9m long. The panels are U-shaped and 300mm deep with the flat back of the U, the exterior face. The 300mm depth of the precast façade is supported every 3m by galvanised square hollow steel columns and internally insulated and clad with a white plasterboard coated metal liner. The windows slots may appear as a random pattern but they have been arranged to allow natural light into the building yet obstruct views into it. We had originally wanted to build the wall using in situ concrete to create a monolithic appearance, but felt we could not achieve the quality desired. Precast on the other hand would give a consistently good finish.

The other striking feature of the architecture is the grey panelled strong box designed for the Command Centre which sits behind the main entrance. Whilst it is not visible from the front, it is expressed on the rear elevation. In this windowless enclosure 60 people staring sit at banks of computer screens displayed on each of their desks. The desks are arranged theatre style in curving rows facing a giant computer screen on the wall. This is the room where staff are watching and monitoring data transactions that occur 24 hours a day, the nerve centre of the complex. It is protected by large precast concrete panels, each one 7m high by 3m wide and weighing 17 tonnes. They wrap themselves around the perimeter of the steel support frame, creating an enclosure with a 30m by 12m footprint.

The curvaceous box of the Command Centre has a glowing blue light around the base which gives it a sense of floating or hovering above the floor. The precast panels have been patterned with a grillage of lines to emphasise the sturdiness of the units, to express their curving profile and to hide the joints between panels.

The building is a steel framed shell with a truss supported roof canopy, with a strong wall elevation to the front and conventional precast elevations to the rear and the sides. Within the building sits the Command Centre and three banks of enclosed storage rooms called Network Rooms standing in a row, each one 28m wide by 40m long by 7m high, separated by access corridors 3m wide. Running along the front of the building is a 9m wide perimeter space for conventional office use, meeting rooms, conference areas and a small café lounge with views through the slotted windows. This area is screened by full-height glass panelling so that the space is bright and visible yet separate from the windowless walls of the Network Rooms. These data storage rooms are protected by fire walls and are clad in heavy duty precast panels. In the data rooms there are a number of robotic machines protected by security cages which spin at very high speed to remove and retrieve data cards from various data banks in an instant. The finished floor of the building is a raised access floor 1m high to feed all the cabling and services required for the data storage installations and the back-up and standby plant.

Upper: Stack-bonded precast units
Lower: Grey/black acid-etch finish

South elevation showing Command Centre and glazed office area

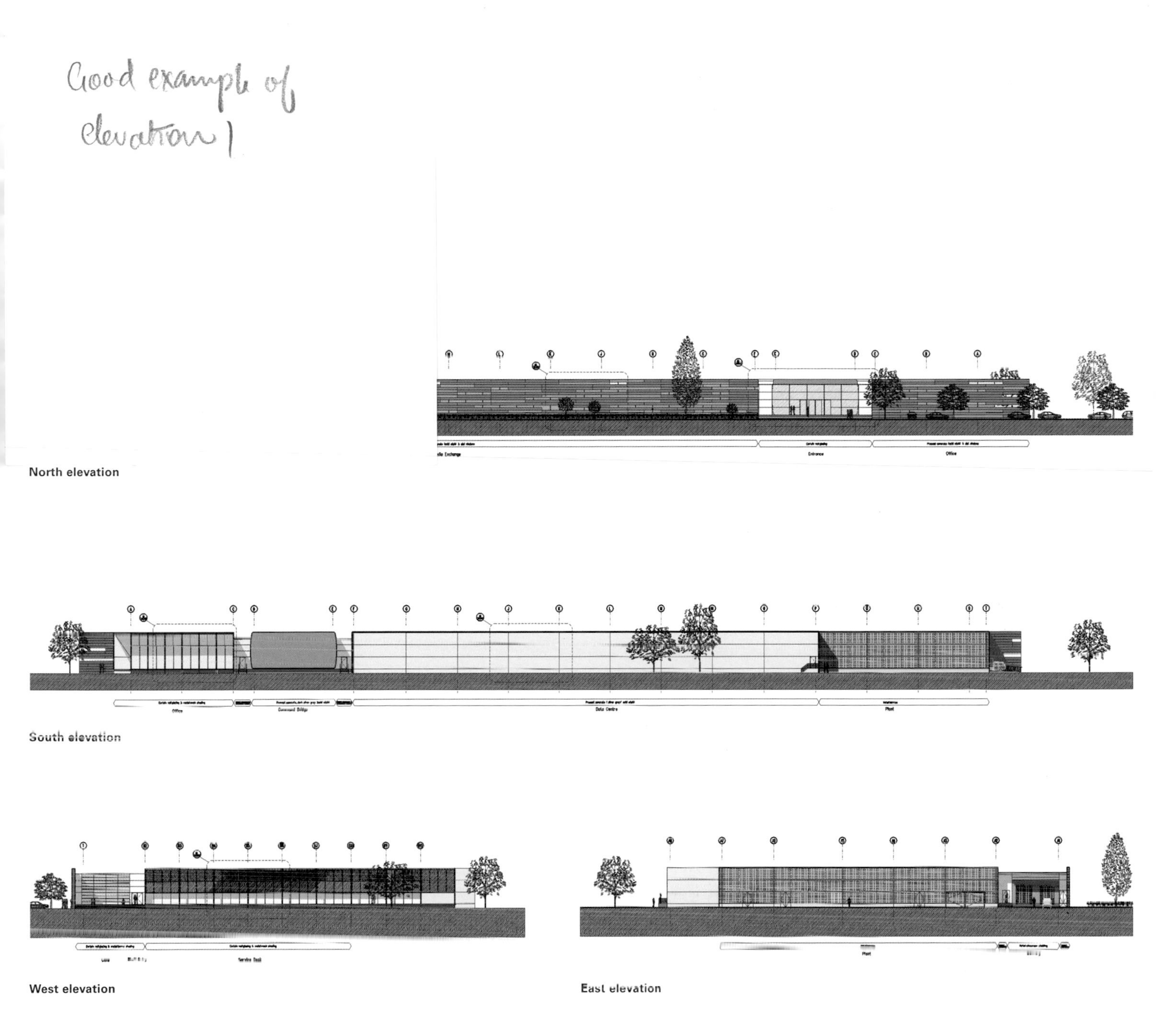

North elevation

South elevation

West elevation

East elevation

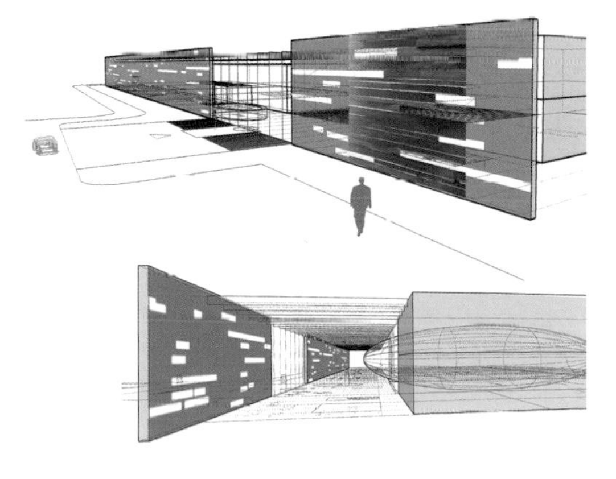

Strong wall interior and exterior views

South elevation with white precast sandwich panels in foreground

In thinking about the aesthetic of the long thin precast panels we tried to find ways to avoid the steel columns running through the window slots but this was in conflict with the preferred positions of the slots relative to the concrete panels. In the end the positioning was felt to be to more important than reading the structure line. A metal clad system was ruled out early on because it was not secure or robust. The grey concrete colour we chose echoes the grey limestone of the region. At first we felt that it should all be the same colour and surface texture and changed our minds when we saw the mock-up in Trent Concretes yard. It lacked variety. What was needed was some diversity in surface texture and sparkle to make it look much warmer and more natural. The different acid-etch to the surface makes the mica in the panels sparkle especially when floodlit at night.

MIX CONSTITUENTS

Sandwich panels rear elevation:
White cement, Clee Hill coarse aggregates, cornish white sand fines

Command Centre and front wall:
Blend of white cement and grey ordinary Portland cement, Clee Hill black basalt coarse aggregate and fine aggregates, black oxide pigment

Mix proportions (per m^3):
Cement 400kg, coarse aggregates 1,320kg, fine aggregates 520kg; water/cement ratio 0.4

Precast Production

David Walker, Trent Concrete

There were three different precast elements supplied for the Experian Centre. One was the feature panel wall on the front elevation, the other the massive sections for the Command Centre and then there were the white insulated sandwich panels for the Network Rooms. We made several samples for the front wall and they came with different surface finishes – acid-etch, sand blast, water jet and so forth. They were cast with the same mix and had two different etches – a light etch which gave a mid-grey surface and medium etch which gave a darker texture and showed more of the mica particles on the surface.

The U-shaped panels were stack bonded with a shadow gap and back fixed to the columns using bolted connections. In all there were 340 panels to fix. The most complicated elements we had to precast were the two-storey high Command Centre panels which had a curvature in both planes. They had false joint lines formed on the face to replicate large ashlar panels. They were very heavy and we had to use a crawler crane to lift them into position. There were 31 panels made typically 3m long 7m high and 150 mm thick which were cast with the same mix as the front wall and given a light acid-etch.

For the rear elevation we used a white concrete mix with a light acid-etch to produce a white reconstructed stone finish to both the inner and outer skin of the sandwich panels. The panels were 9m long, 2.37m high and 315mm thick. The outer wall was 65mm, the inner load-bearing wall 165mm thick with insulation between them 75mm thick. We supplied 104 units to the site each one weighing 13.6 tonnes.

We use marine ply formwork to make the moulds and then line the contact face with GRP which we sand-smooth to give a fine smooth finish to the concrete. We achieve at least 30 casts out the mould. In the past we used to coat the ply with polyurethane paint and rub it smooth but that did not allow as many castings as the GRP, before we had to re-apply the coat. We always prefer to cast long panels which means less fixing time on site and lower installation cost.

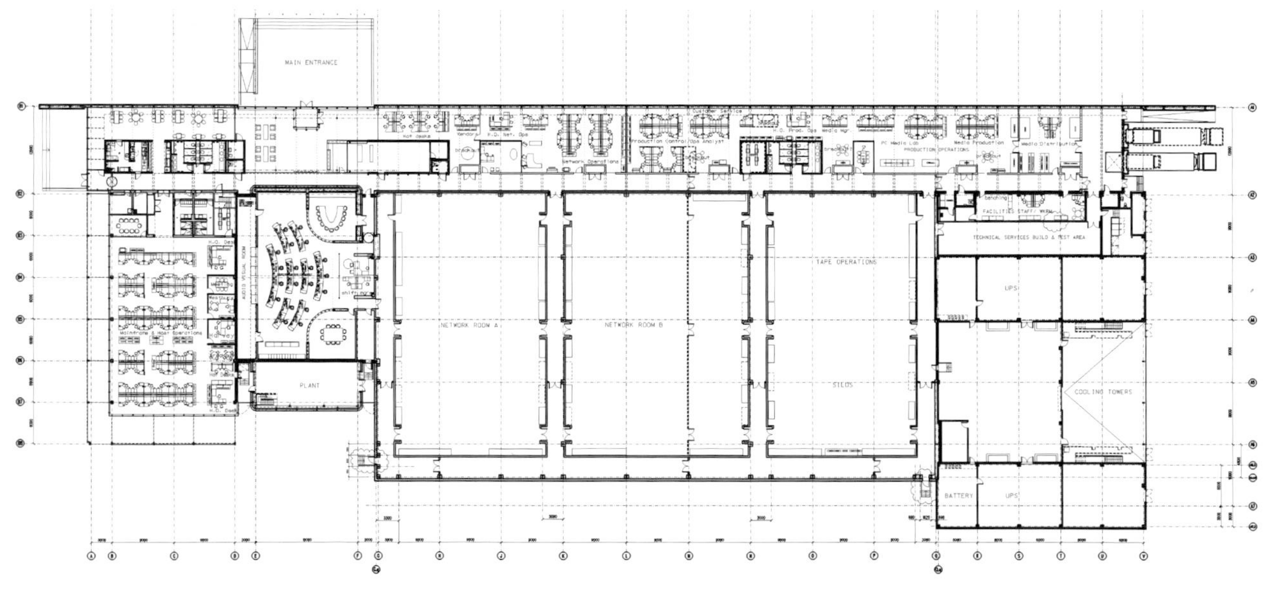

Floor plan

PROJECT DATA

Client: Experian Ltd
Architect: Sheppard Robson
Structural Engineer: BWB Partnership
Services Consultant: Troup Bywater & Anders
Main Contractor: Bowman & Kirkland Ltd
Precast Manufacturer: Trent Concrete Ltd
Completion: 2003
Contract Duration: 2 years
Construction Cost: 25 million GBP
Cost/m^2 precast units: Front wall 263 GBP, command centre wall 306 GBP, sandwich panel rear elevation 274 GBP

Building section with strong wall and internal corridor area

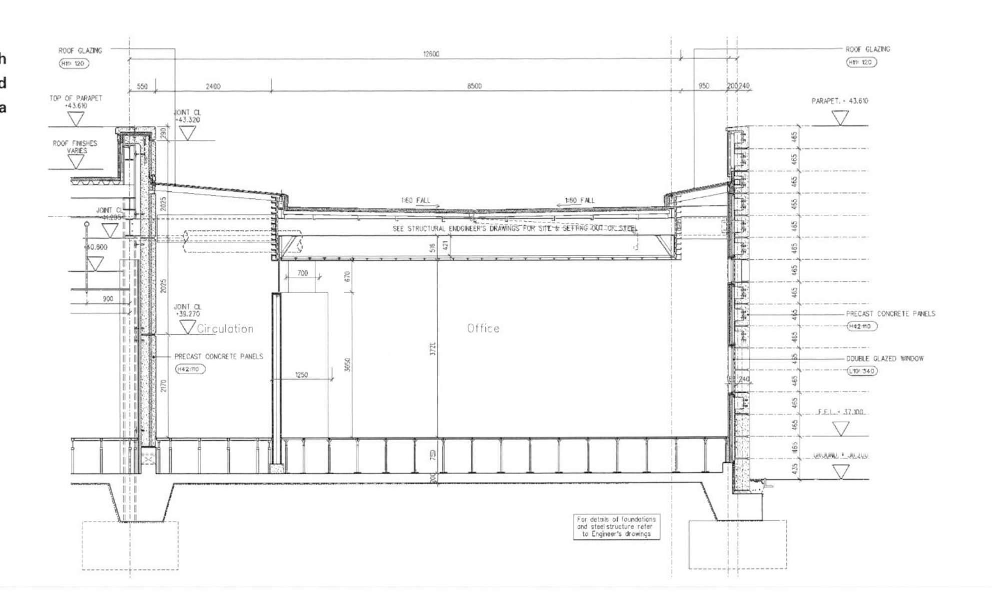

Top: Reception
Bottom: Interior façade of strong wall

Top: Internal corridor and floor lighting to Command Centre
Bottom: Visitors area

SCOTTISH PARLIAMENT BUILDING, EDINBURGH

EMBT Barcelona and RMJM Architects

Location

The Scottish Parliament Building sits on a four acre site one kilometre east of the city centre, at the foot of Edinburgh's historical Royal Mile, adjacent to the Palace of Holyrood and in the shadow of Salisbury Crags.

Architectural Statement

Enric Miralles and his Barcelona practice EMBT joined forces with RMJM architects – a well established Scottish practice – to enter the Scottish Parliament design competition, which they duly won in July 1998. The plan was for Enric Miralles and EMBT to lead the concept design and for the 40 strong RMJM team to carry out the detailed design, issue contract drawings and oversee the construction.

The architectural objective was to create a unique institutional building that was open, anti-classical and non-hierarchical, that embraces the landscape defying all the canonical rules of architectural composition.

The single most dramatic idea was that the building should sit in the land and align itself symbiotically with it. Formally this is symbolised by the landscape sweeping in from the Salisbury Crags up to and under the raised forms of the superstructure. The relationship of the site to the ridge of the inspiring volcanic landscape was crucial in defining the design. Equally important were the close links with the ancient parts of the city; for example the public entrance and the Debating Chamber look out towards Holyrood Park and the Palace of Holyrood and Dynamic Earth beyond.

MSP building

The building plan is basically a loosely arranged U-shaped cloister opening up to the landscape and containing it like a cupped hand. The enclosing western, northern and eastern edges form the boundary in the conventional sense, as dynamic urban set pieces. The complex is split into three main building types and functions. Firstly there is the massive six-storey MSP Building for the Parliamentary Members, which defines the western edge of the site and completes the distinctive rigg plan of the medieval city. Secondly there is Queensberry House to the north, a grade A listed building that existed on the site from the early half of the 17th century. Continuing along the northern boundary adjoining the Royal Mile you find Canongate, the formal entrance to the MSP Building. To the east of it and to the southern fringes stands a cluster of tower structures that envelop the Debating Chamber. The four Tower buildings that nestle round the Debating Chamber house the Committee Rooms, the various support staff and the Media Centre. The Debating Chamber and the public concourse lie on the eastern fringe of the site. The Debating Chamber is positioned to face Holyrood Palace with the building scaling down on its eastern flanks in sympathy with the Palace. There was a desire to anchor Canongate in the streetscape with a great muscular wall and for the buildings to be loosely arranged around a cloister garden in the manner of a university campus.

The Debating Chamber itself has been designed as a theatre with 40% of the seating accessible to wheelchairs. The focus of the chamber is the huge truss span roof made from laminated oak with tension ties and connecting nodes of stainless steel. From the outset the seating layout of the chamber was to be homogenous where party boundaries were to be invisible to discourage the confrontation embodied in the Westminster model.

The Canongate Wall is the massive elevation along the High Street and the continuation of the Royal Mile. It commemorates the presence through history of the Parliament of the Old City by a collage of vignettes, memories and quotations from Scottish literature. The association with the landscape is symbolised by pieces of diverse stone representative of the geological strata of the land. The Canongate building itself is an engineering tour de force where a 16m cantilever, supported on two vierendeel trusses, converge the elevations to meet at a sharp pointed edge.

Canongate wall detail

The Canongate Wall

Kevin Grubb, RMJM Architects

Blast requirements ruled out the use of conventional windows to the wall. It was always the intention that it would be modelled and decorated in some form. An early idea was a 'constituency' wall, with each constituency in Scotland to be represented. Over the design period, this crystallized into the current proposals.

Canongate wall

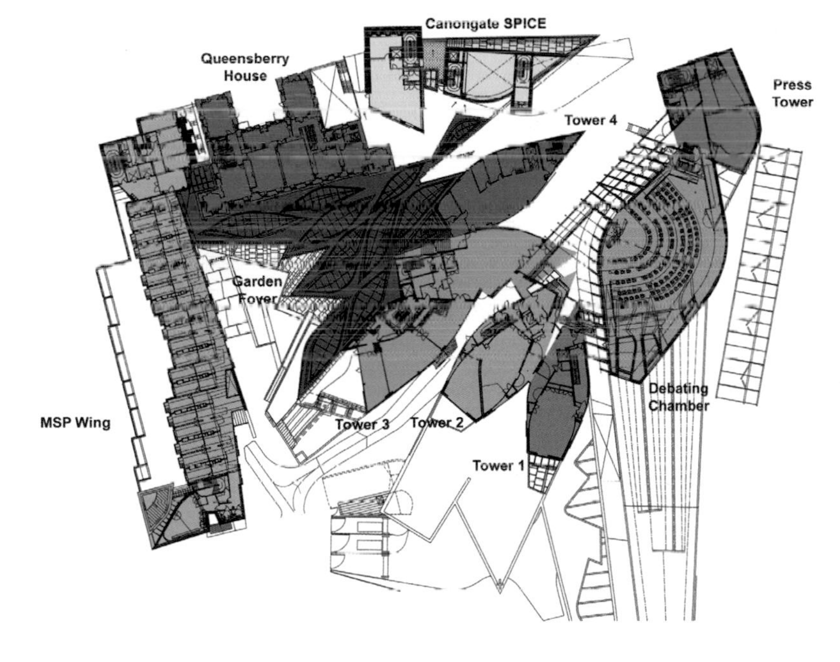

Site plan

Public entrance

Reading the wall from left to right, the first element we encounter is a Miralles sketch of the Edinburgh skyline, cast into the rough sandblasted concrete. Then follows a collection of craggy stones, post fixed into recesses cast into the wall as though they have burst through the surface of the concrete. Throughout the design process, we worked to bring the landscape of the crags through as far northward into the site as possible. These craggy stones are the culmination of that process, as Salisbury Crags crashes right through the building to arrive, defiantly and proudly, in the city. Moving westwards again up the wall, the main area is finished in smooth fair faced concrete into which are fashioned a series of 'windows', not glazed but formed in niches cast into the concrete.

During the design process a 1m square sample of each element was produced to check the design feasibility and to refine the methods employed to produce the panels. After the problems had been resolved the elements were combined into a full scale mock-up to test if they could be cast into the curved forms required. The knowledge gained from this process allowed us to develop methods for communicating the complex design information onto CAD drawings to issue to the precast contractor.

The main geometric elements consist of a top and base line of gently curved panels, a middle region of flat panels with insert stones and a series of vertical fins. Both in plan and elevation these elements do not run parallel to each other producing a subtle distortion. Due to casting tolerances the actual concrete panels always varied slightly from the theoretically designed panels in both size and geometry. With the complex relationship between the panels these slight variations tended to accumulate over the length of the wall resulting in misalignment and panels not fitting correctly into position. This was alleviated at the design stages by building in tolerances in the joints between panels and by surveys of the existing and surrounding structures. The 3D model of the entire wall was utilised to produce a comprehensive package of setting out information and to check the progress of the panels during installation. The individual concrete panels were craned into position and fixed back onto the main in situ concrete structure via stainless steel framing. The steel frames allowed the positions of the panels to be adjusted to take up any tolerances.

Precast Concrete

David Shillito, Malling Products and Gary Lucas, Patterns and Moulds

Malling Products are a subsidiary company within Laing O'Rourke, who built the concrete frame. The finished quality of the elements had to be good and we therefore only resorted to in situ concrete construction where it was not practical or possible to precast.

Many stories and rumours abound about the design ideas of the Canongate Wall and the bamboo motifs on the MSP boundary wall. Enric Miralles had given some hint of his intentions on rough sketches. We were given one drawing for the Canongate Wall from which to evolve the precast panels and give a price. The drawing was a combination of a rough sketch, a montage of images that had been cut out of magazines or from photos taken of the surrounding landscape... and some poetry! That was all. By no means the clarity and precision we have come to expect from architect drawings.

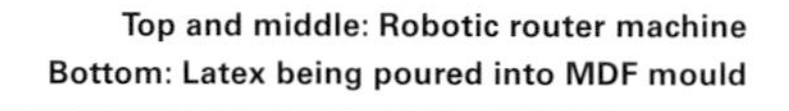

Top and middle: Robotic router machine
Bottom: Latex being poured into MDF mould

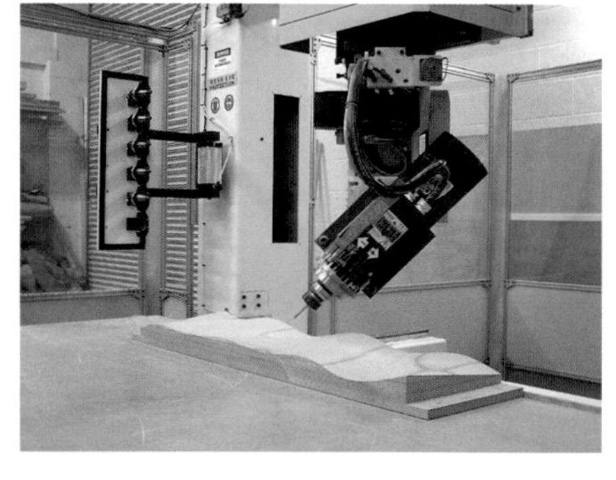

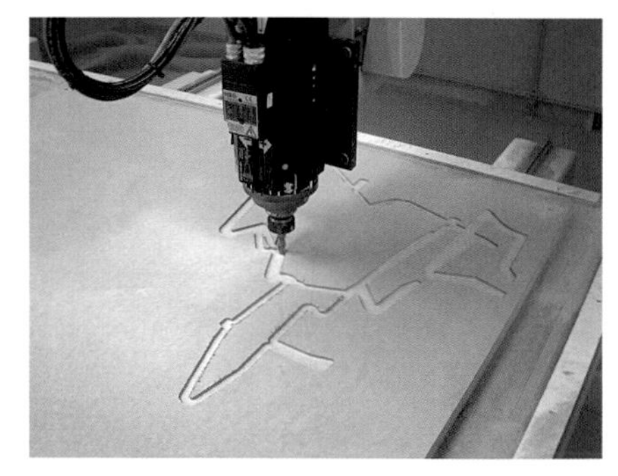

We were already working with Patterns and Moulds trying to find an economical way to make the moulds for the unique bamboo pattern relief on the flat panels for the MSP boundary walls. The challenge was not how we cast or finished the concrete but how we could make bespoke moulds which were curved along an asymmetric axis and indented with large and small pockets in a random order to receive pieces of natural stone. Using conventional methods of mould making and employing teams of carpenters it would have taken five times longer than the robotic router method we invested in. The cost of the finished concrete using the robotic router was around 4,600 GBP/m^2 for these panels. It would have cost at least 10,000 GBP/m^2 using traditional methods and taken at least three weeks to make each master mould.

When the sketch for the Canongate Wall was further developed by RMJM into a series of CAD models, we knew that the master mould could be made using upgraded software to drive a CNC machine. The actual precast panels to be made were each 3m by 4.5m in size, but as we can only tool panels sizes of up 3m by 1.5m we had to split the 3D image into thirds, in order to generate the information required to tool the whole panel. To programme the CNC machine takes a technician about four days sitting in front of a computer.

The biggest bug of the system is the quality and accuracy of the Auto CAD drawing. The information input for a CNC machine works to an accuracy of five decimal places, so if adjoining lines on the CAD image is out by less than half a millimetre the machine will not function and throw up an error. The software produces the tool path to drive the cutter and sends positioning code signals similar to a stereoscopic image.

A 50mm deep, 3m by 4.5m MDF panel is laid out below the router head, cutting 25mm into the MDF to form the relief patterns. The machine cuts a 3m

Canongate SPICE Building (Scottish Parliament Information Centre and Information Technology)

Canongate elevation

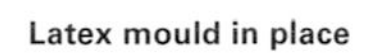

Latex mould in place

Casting the leading edge panel

Inserting the cut bricks and stone pieces

Canongate wall and CNC cut patterns

by 1.5m section in 14 hours and it can take up to 1.5 days to complete the panel. The cutting head rotates at 25,000rpm and the cutting tool is chai the course of the process.The first stage is the rough cut using 16-25mn diameters, when a large amount of material is removed. When the rou down to the finishing stages, the machine makes hundreds of passes c panel to bring the cut surface to a smooth finish, using ball-end cutter h

Motif on Service Bldg!

The MDF is then waterproofed with varnish and set in a boundar The latex rubber – a thermosetting two-part polymer resin – is poured MDF. It has the consistency of treacle and fills all the depressions and c level across the boundary frame. When it has hardened it is lifted out to us at Malling Products in Grays. The latex rubber mould is placed in th mould frame and the release agent applied. The stones that appear in the and windows in the panels are then placed in position. Some of the sto recovered from the old brewery that once occupied the site. Every piec had to be carefully cut to size and shape and placed in the correct rec panel before concreting. The Scottish Parliament concrete mix was pc the mould and left to harden. The mix was a pale grey concrete using ordinary Portland cement and crushed Derbyshire limestone fine and coarse aggregates.

Canongate panel in position

The 250mm thick precast panel which was reinforced with steel mesh, was left to cure in the mould for 24 hours. When the panel was removed some surfaces had to be sand blasted and others left with a lightly polished finish. We had to mask all the fair-face non blasted surfaces with plastic insulating tape and fill up the stone rebates with weak mortar so that during blasting it would not get eroded. That was the most difficult part of the precasting work.

An important issue on the Scottish Parliament Building concerning all precast work is the surface quality and preciseness of the panels. Many of the panels had patterns on the surface that were continued onto adjacent units. If we worked to the tolerances we are allowed to use (i.e. +/-10mm for fixing tolerance, plus +/-6mm for manufacturing tolerances which is collective of +/-12mm,) we can have ugly edges of discontinuity as wide as 15mm to 20mm. With the CNC routed moulds we achieved a manufacturing accuracy to within one millimetre of exactness. Although the panels on the Canongate Wall were heavy, up to 18 tonnes, and there was deflection in the cantilever steel supporting frame, we ended with a positional tolerance of only +/-2mm which is phenomenal.

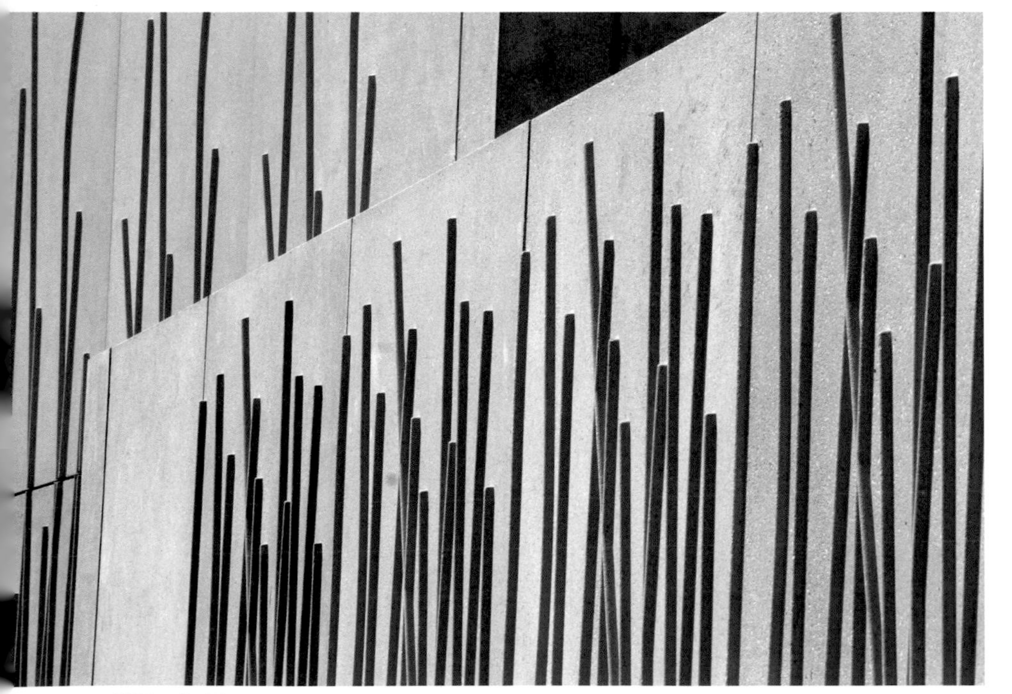

MSP wall with bamboo motif

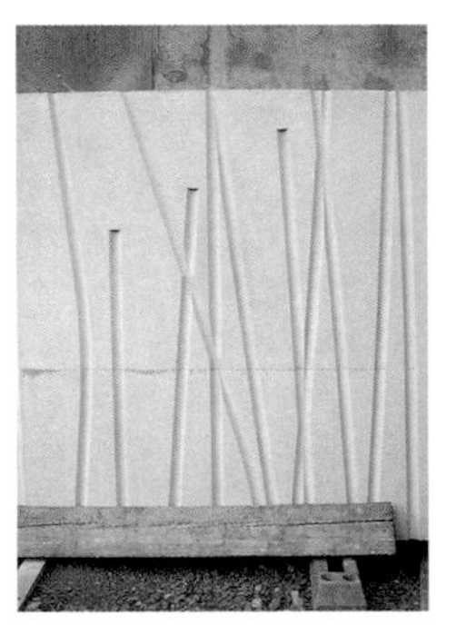

Panels in precast yard

PROJECT DATA

Client: Scottish Parliament Corporate Body
Architect: EMBT/RMJM Joint Venture
Structural Engineer: Arup
Services Consultant: RMJM Scotland
Construction Manager: Bovis Lend Lease
Concrete Structure: Laing O'Rourke
Precast Elements: Malling Products
Completion: 2004
Contract Period: 5 years

Canongate wall showing separation
of upper and lower curvature

ARLANDA CONTROL TOWER, STOCKHOLM

Wingårdh Architects

Location

The Stockholm/Arlanda Airport is located about 40km north of Stockholm in the county of Sigtuna. There is a fast train service from Arlanda to Stockholm City, and a bus between the airport and City Terminal in central Stockholm, near the Central Station.

Architectural Statement

Gert Wingårdh

In the 1990s Arlanda International Airport just outside Stockholm planned for a considerable expansion, with new runways, terminals and a new flight control tower whose size and central position was to be an emblem of the new airport.

The composition of the tower is based on a classical tripartite division into base, shaft and capital. The design builds on the theme of juxtaposing polar opposites. The tower is striped black and white, like a lighthouse or cairn on the seacoast. On top are two rooms for directing traffic, the upper one for monitoring take-offs, landings and the surrounding airspace, the lower one for monitoring aircraft at the gates and taxiing them towards runways. The scheme clearly differentiates between technical equipment and the human workforce, separating the two and cloaking them in black and white respectively. The theme recurs in the facilities housed within the base of the tower.

For flight control, the tower is only a means to provide a clear view. The control room is the building's raison d'être and therefore its most dominant element, while the shaft is kept as slim as possible. From the upper floor of the control tower the controllers watch over the air space and runways, from the lower floor they survey the taxiing areas and the aircraft docking. This separation of functions forms the basis of the tower's design. The tower is a double shaft, one is black for the upper floor, the other white for the lower floor, while the base building is glass and white on one side and black on the other. The repeating narrow rings around the two tower shafts are decorated with citations from Antoine de Saint-Exupéry's 'Postal Flight South', a proposal suggested by artist Silja Rantanen which we emboldened to give the stripes the necessary significance they would otherwise have lacked.

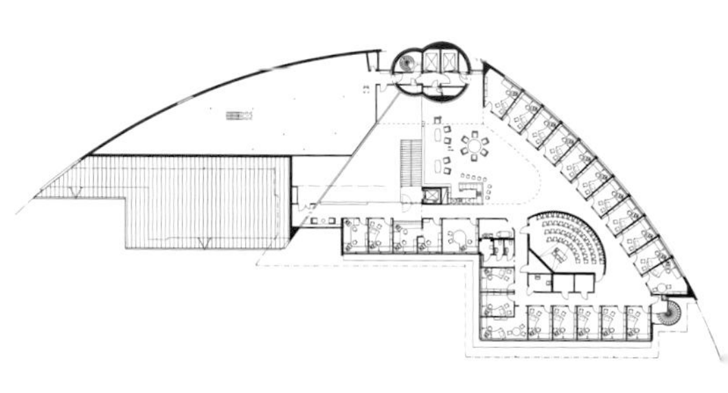

Ground floor level

Control tower and aircraft parking apron

The base building is more than just a technical utility. The architectural task called for a formal expression going beyond what was technically required. As in many other works, we like to show our appreciation of mid 20th century art and architecture. The form and pattern of the buildings resemble op art, and to an even greater extent the ceramic art in which Swedish artisans like Stig Lindberg designed objects with black-and-white decorations.

Precast Construction

Arne Hellstrom, Strängbetong

The double cylindrical shafts of the tower were cast in situ, using a slip form construction. The façade has a single skin of precast concrete with a polished finish, pigmented black for one cylinder and white for the other. One core contains the lifts and service runs, the other the staircase for the fire escape. The precast panels were 90mm thick and 1.2m high and cast in sections that were quarter arcs of the cylindrical circumference. They were placed starting from the top of the tower downwards, not from the base up as you might suppose. We had a special crane mounted on the top of the tower, which lifted every precast element into position. For the tower structure there were two types of precast panels. There are platform

Precast factory casting shed

Installing the curved panels
for the tower

Panel being hoisted onto the slip formed core

Black precast for the upper and
white for the lower control room

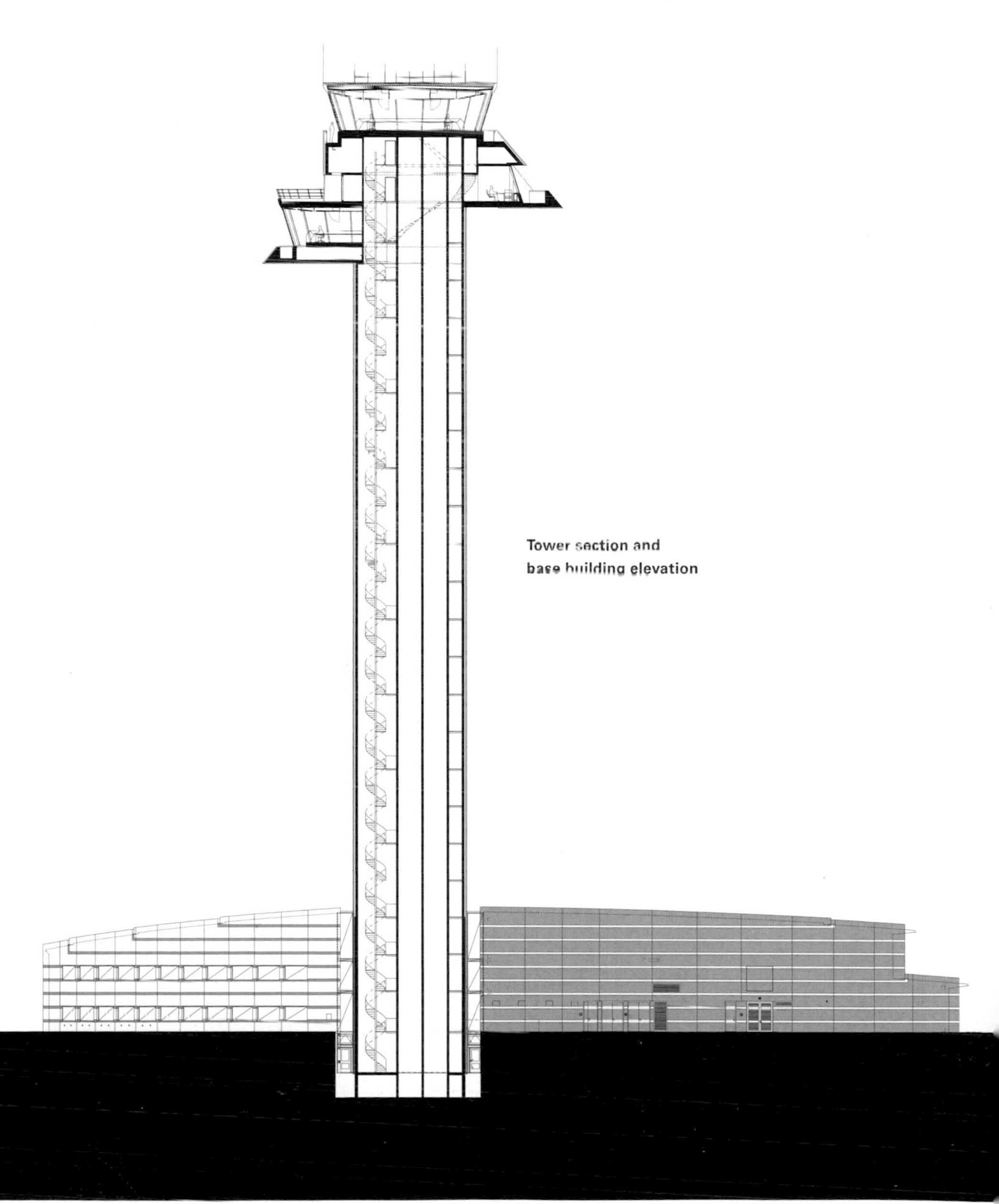

Tower section and
base building elevation

panels that clad the underside of the two saucer-shaped platforms for the air traffic and plane parking controller rooms which have a unique geometry, thus no two units have the same curvature. To ensure a perfect fit the panels were match cast so their abutting edges were an exact fit. These panels have special connections to tie them to the steelwork of the cantilevered platform. For the double shafted core, the panels were attached to the concrete tower using metal anchors with an air gap between them and the structural face.

The black concrete units were made using grey cement, black marble fine and coarse aggregates and black pigment. An additive creating 5% air is mixed into the concrete. The white concrete was made from white cement, titanium dioxide (white) pigment and white marble aggregates. The concrete for the flat panels were cast in wooden moulds and left for one day before they were lifted out. The cylindrical panels were cast in metal moulds lined with a plywood face and left to harden for about three days before they were polished.

The exposed concrete surface is wet-polished, first with coarse diamond headed discs and progressively with finer ones until it is rendered smooth. The air holes on the surface are then filled using fine slurry of white or black pigments and cement. After leaving the slurry to harden for a day or two the surface is again wet-polished with just the fine diamond heads to remove the excess cement paste, providing the surface with a terrazzo-like lustre.

The cylindrical single skin panels for the shafts were cast vertically in moulds. They were turned on their sides with the exterior face uppermost for wet grinding and finishing. We do get a few problems with efflorescence on black concrete finishes. On white concrete finishes, however, it is less noticeable. By wet grinding the surface we tend to reduce the risk of it occurring but it can still prevail. It is less a problem with a single skin of precast where both sides of the panel are exposed to the air and can dry out, and more likely to occur on the surface of the sandwich panels as they can only dry out from the exposed face, so all the water vapour migrates out of that side and this increases the risk of efflorescence on the surface.

The base building which is on two floors, has external load-bearing sandwich panels as the perimeter floor support system. The black and white polished exterior panels of the sandwich panel are separated from the internal load-bearing panel by 150mm of insulation. The façade panel is 80mm thick, and the internal load-bearing element is 120mm thick. This is a common construction method in Scandinavian countries. The inner wall is fixed to the in situ concrete frame. The outer and inner layers are connected by stainless steel connections. Most sandwich elements are 3.3m high and 7.2m long.

The sandwich panels are cast on tilting tables, the face coloured concrete mix is placed in the mould first, the insulation then covers the wet concrete before placing the rebar and pouring in the backing mix of ordinary grey concrete for the load-bearing element. The day after the casting, the tables are tilted up and the precast unit is lifted out by an overhead crane and taken to the finishing area for wet grinding. When that is complete the panel is taken to the stockyard and stacked on wooden sleepers before they are transported to the site.

There were 970 precast elements supplied to the contract which took six months to install, including working through an unusually cold winter.

Views of base building façade

Interior of tower and entrance to lifts

PROJECT DATA

Client: Luftfartsverket
Architect: Wingårdh Arkitektkontor AB
Structural Engineer: Luftfartsverket Teknik
Services Engineer: SCC VVS-teknik
Main Contractor: PEAB Sverige AB
Precast Manufacturer: Strängbetong
Completion: 2001
Construction Time: 2 years

View of double cylinder tower

HENRY DUNKERS CULTURAL CENTRE, HELSINBORG

Kim Utzon Architects

Location

Situated on the newly developed Northern Harbour front of Helsinborg, the Cultural Centre is in the centre of the town and close to the bus terminus. Its row of periscope roof lights are unmistakable on the coastal skyline. Helsinborg is an attractive city with panoramic views across the Øresund towards Denmark and a popular destination for tourists.

Architectural Statement

The Centre constitutes the southern extremity of the major regeneration plan for the former industrial port Norra Hamnen (the North Harbour). Newly-built blocks of flats with cafés on the ground floor stretch along a beautiful quayside walkway, part of a vibrant new urban district tucked in between blocks of 19th century flats and the new marina on the Øresund.

The Dunkers Culture Centre serves as a link between the city and the sea, a function accentuated by an 18m broad passage the full length of the complex. Via an access square, it connects the city to the quayside, the marina and the magnificent landscape of Øresund. The Centre consists of a differentiated building complex comprising juxtaposed parts connected by the indoor street. It is a city within the city and serves as a living room for the citizens of Helsingborg. The broad passage divides and draws together the Centre. The main entrance is in the east, off the access square, where the street and a wide stairway channel visitors upwards to a variety of cultural events and activities in the four-storey complex. Corridors over an open sculpture yard link the entrance to a foyer facing the west, where the theatre, concert hall and restaurant are located.

The various parts of the building share a constructive system as well as the vocabulary consisting of prefabricated columns, beams and brick-fillings. The working and teaching areas are relatively simple four-storey buildings, which border on the culture block on the north, east and south. The working areas encase the exhibition buildings, which are situated to the north and south of the central broad east/west passage. The exhibition spaces are spread over two floors with double-height rooms. The component parts of the building system on the second floor form a simple and flexible space that meets the requirements of art exhibitions for a combination of north-facing studio lighting and artificial light. The arched roofs of the theatre and concert halls surge like sculptural waves along the waterfront.

Discussion

Peter Forssman, Skanska

The Cultural Centre is located on the transformed Northern Harbour front which until a decade ago was an industrial wasteland covered by railway tracks, crawler cranes, storage sheds and silos. It was designed to be a dynamic forum for encouraging diversity in art, history, drama, dance and music with the intention of promoting an enriching dialogue between the public and the artistic community.

It is not as though Helsinborg has never had an arts centre, this is the third one to be built in the past three decades. Many impartial observers argue that Helsinborg has more art and cultural centres per capita than any other city in Europe thanks to the bequest left by the industrialist Henry Dunkers who made his fortune manufacturing rubber footwear, world famous tennis balls endorsed by Swedish tennis legend Björn Borg and galoshes. He died childless in 1962 and donated his great wealth to the city to promote art and cultural pursuits.

In 1996 the city ran an international competition to design a new arts centre. The winning entry by Utzon Architects, 'The City By the Sea' was based on the medieval town, which can be read as a collection of houses connected by numerous walkways, open terraces, and luminous courtyards at different levels. The city within a city structure is underscored by the intricate layout of the building which comprises numerous boxed spaces in the centre of the floor area, each having its own expressive form. Perception of the building changes as you walk through it, appearing enclosed towards the centre as you enter the exhibition halls and galleries, and open and transparent along and through the spacious periphery corridors.

Utzon solved the familiar problem of maintaining contact between town and harbour areas by opening up the ground floor so that you can see right

'The City By the Sea' waterfront

Lantern skylights and curved roof of concert hall

through the building out to the sea on one end and to the town square on the other. To reinforce this feeling the ground level of the building is raised 1.5m above the quayside to create the illusion of the sea reaching the edge of the building when seen from the town square entrance.

The Centre has a total floor area of 16,000m^2 spread over three levels making it the largest building of its kind in Scandinavia. It houses a vast art gallery on the third floor and smaller ones which can double as exhibition spaces on the ground floor. The concert hall and theatre building can be distinguished by the exaggerated curve of the steeply sloping roof pods. They appear to be two independent buildings with high impressive gables and zinc covered roofs, designed to represent two huge waves sweeping in from the sea. On the third floor central to the complex, is the main art and exhibition area of 650m^2 with its array of egg box roof domes.

Upper gallery space

The numerous different functions of the centre are reflected in the widely varied structure of the façades and roofs. The building has an urban character on the town side, while facing the Øresund it opens out on to a generous courtyard and the façades have more sculptural emphasis. On three sides there are rectangular dormitory buildings with conventional mono-pitch roofs and conventional storey-high floors housing the administrative offices, workshops and music school. They surround the art galleries and exhibition halls which are characterised by rows of roof lights which look like giant ventilation stacks.

Construction of 'egg box' light well

Completed lantern light well

Concert hall

Upper exhibition hall

Construction of perimeter colonnade

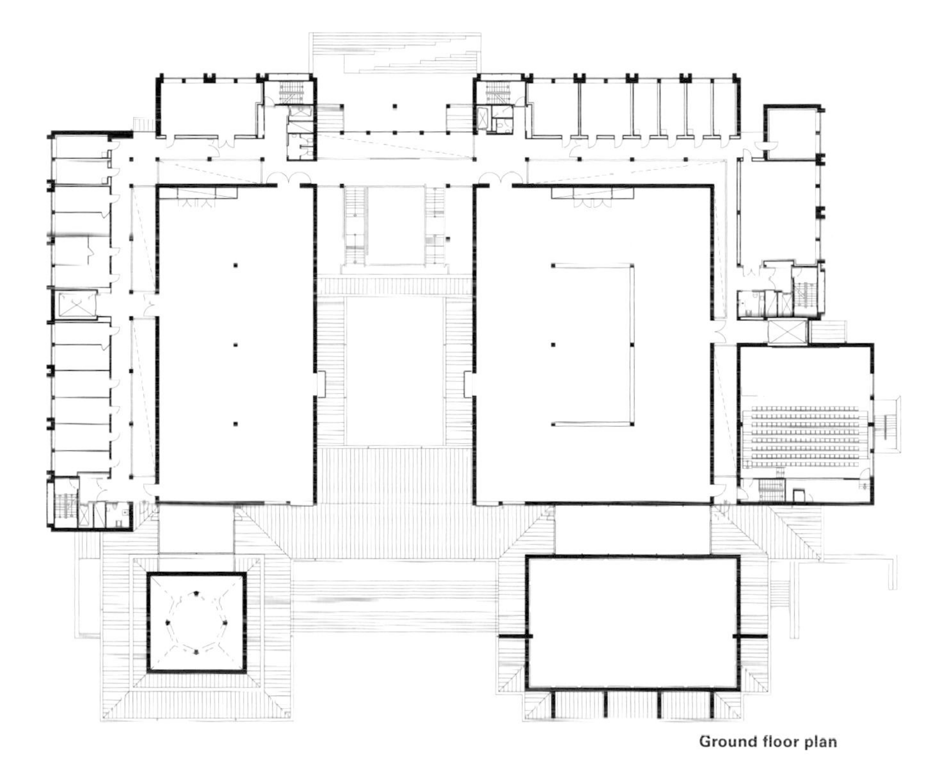
Ground floor plan

Staircase

The concert hall has large deeply recessed windows with automatic shutters, so that audiences can enjoy the views out to sea before and after concert performances. The ground floor restaurant and bistro also overlooks the quay with balcony seating on ground and first floor. The roof over the restaurant is crowned by an octagonal pavilion framed in precast concrete that rises up from the first floor. It is a modern sundial with narrow windows in the cupola that admit strips of light marking the sun's course.

The exposed frame and floors of the building are raw precast concrete elements painted white. The paint is a silicate-based 'Keim' paint with white pigments which bonds chemically to concrete and brickwork substrates. It consists of potassium silicate binder with inorganic fillers and natural earth oxide colours. This solution flows into the pores of the concrete substrate where it slowly crystallises. Thus it creates an extremely durable, waterproof surface while remaining vapour permeable so that moisture in the substrate can escape into the atmosphere. The façades are a combination of brickwork and precast panels painted white. The restaurant walls and the sundial pavilion are clad in white glazed tiles. The complicated roofs are covered in zinc cladding as are the distinctive eaves gutters. The painted T beams in the ceilings alternate with wood-wool acoustic panels. The acoustic panels are made of cement-bonded fine wood wool which offers good insulation value and is fire rated. The vast convex ceiling void created by curved glulam beams in the concert hall and theatres is covered with bleached pine panels.

Section

Main corridor and exit leading to waterfront

PROJECT DATA

Client: Helsingborgs Stad Kärnfastigheter
Architect: Kim Utzon Architects
Structural Engineer: Tyréns
Main Contractor: Skanska Sverige AB
Precast Manufacturer: Skanska Prefab
Completion: 2002

Octagonal cupola with lantern slots above restaurant

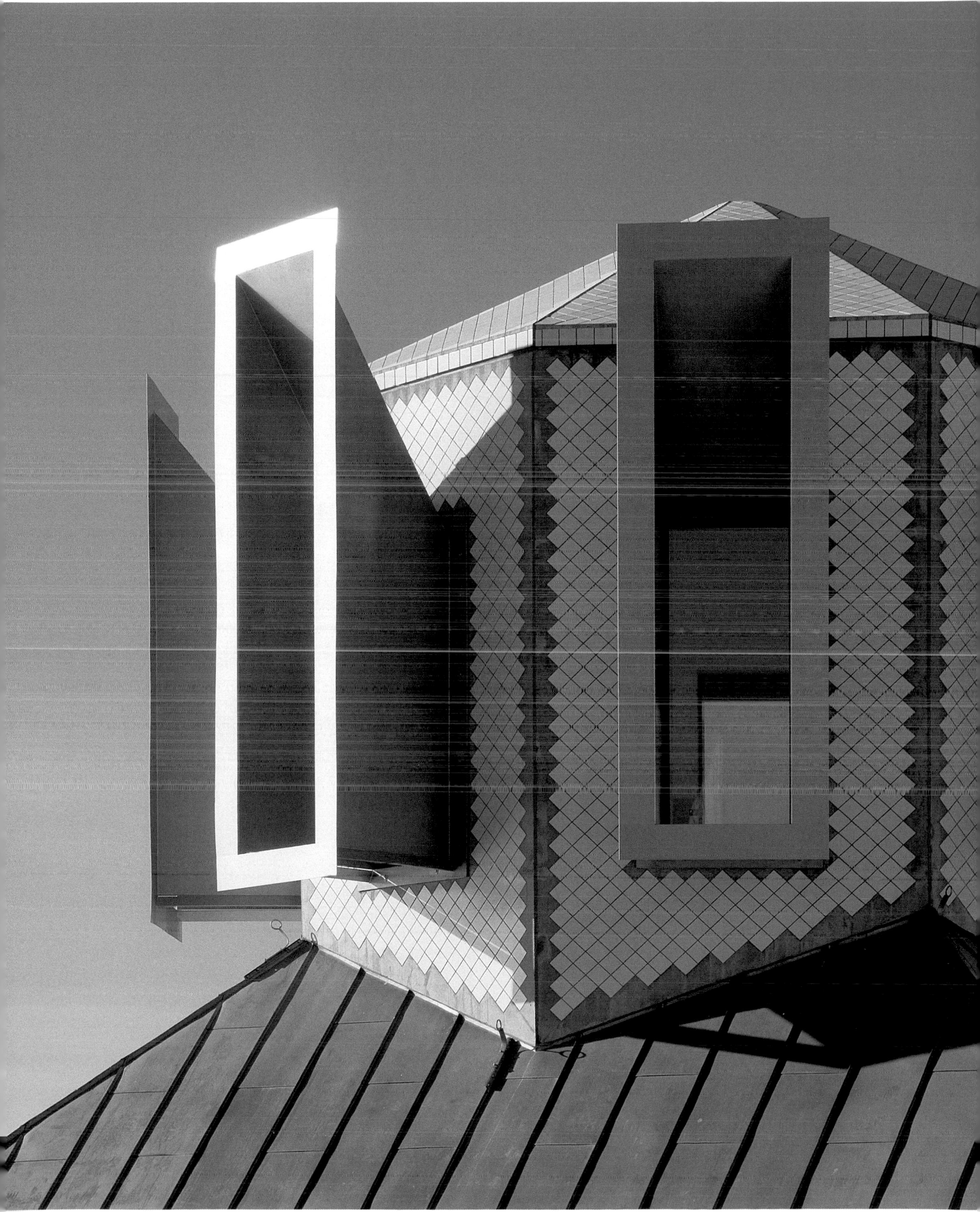

KATSON BUILDING, STOCKHOLM

White Architects

Location

The Katson Building is situated about a ten minute ride by metro from the city centre. Take the Green Line Metro from T Centralen station and go to Skanstull on the southbound route. On leaving the metro station head towards the Skanstull Bridge and the lock gate of the canal below it. The building is adjacent to the lock gate.

Architectural Statement

The new office building lies in a spectacular urban landscape where the Skanstull Bridge extends over the building and, together with the Hammarby lock, forms the backdrop to the site. It is part of the plan for the Hammarby dock expansion, a new residential precinct in what was previously an industrial dockland of Stockholm.

The building has an uncomplicated external form – it is a long narrow rectangular glass box with five storeys and a roof. The façade consists of a light metal framing with glazing panels. On the roof is a single storey timber structure set back from the building edge and a terraced landscape. On the quay side, built over the canal basin, is the main entrance to the building.

The strategic position of the building demands that all sides of the building – since they are so visible – are effectively a main elevation. Half of the building projects over the adjacent canal and at dusk the façade is mirrored on the water surface.

The formal language of the architecture is characterised by hard-driven structural demands of exposed concrete soffits, columns and beams, neatly arranged mechanical and electrical systems, and by precise detailing. The building is historically related to industrial buildings of the past with large roof lights and side windows. It has an internal form and plan that is reminiscent of a dockland warehouse.

The project has embraced high environmental standards and despite the glazed façade, low energy consumption has been maintained. The internal climate and temperature are controlled by the use of concrete's thermal mass. Water is fed through plastic pipes laid on top of the precast floor slab, thus increasing the cooling capacity of the building. This energy contribution is free as cool water is pumped out of the canal and re-circulated back again.

Exposed precast floor beams and columns

Discussion

Bengt Svensson and Linda Mattsson

The building is confined on three sides. On two adjacent sides it borders on the junction of the roadway and the high level canal bridge, and on the third one on the canal itself. Theses boundaries fix the perimeter of the building plot and restrict the building height to five floors, to sit below the bridge.

The design concept harks back to the International Style of the 1960s, which may seem retrogressive to some; its timeless qualities, however, continue to have a lot of significance and relevance in architecture today. We wanted to express the structural fabric of the building as the architecture. There was to be no redundancy. The detailing of the building had to be precise. There was no false ceiling to hide any unsightly service ducts, overhead lighting conduit or a poorly finished structural soffit.

View of main façade

The precast concrete structure we have detailed is displayed in a glass showcase. We worked with the structural geometry, to make the joints and connections appear seamless, as though they had been poured. The double T beam soffits were flush with the primary beam soffits. The widths of the primary beams were made the same width as the columns and so on. We chose the natural grey of cement as the colour for the precast with a surface finish that was struck from the moulds. Local cement, sand and aggregates were our structural stone. We did not want a white artificial concrete. There is nothing wrong with grey concrete that you see in functional car park structures in Sweden. It was not our intention to put the precast manufacturer under pressure to produce some dream-like aesthetic finish, with acid washing or polishing. The double T beams which are all exposed, span across 12.65m in the open office area and across 4.15m in the corridor span. They are supported on primary beams that span 7.2m between the three columns runs. The corridor span is the partitioned zone for archive storage, meeting rooms, the lifts and the services core. The columns were cast in double-storey high lengths and were a bit flimsy during erection and had to be temporarily braced until the beams were connected and toppings to the double T slabs were poured.

Left: Open stairwell between upper three floors
Right: Roof top restaurant building

Interior finishes of concrete and wood surfaces

We wanted an open plan transparent building with light-filled interiors that offered views across the canal. The structure was designed to be adaptable for future changes so that installations can be removed and replaced with relative ease. The ducts, the light fittings and sprinkler systems are suspended or attached to inserts cast into the concrete floors. They are contained within the depth of the T beams so that they are only visible from below.

We overcame the lack of dimensional compatibility between the precise lengths of glazing units and the construction tolerances required for the precast elements, by using short adjustable steel beams to support the curtain walling. The glass is double glazed with a special film inserted internally for added security. The floors are covered in forest ash ply, the doors are forest ash, and the walls in the corridor zone are concrete walls and stud walls panelled with abache, perforated for acoustic damping. The main staircase is precast concrete and the internal staircase is folded plate steel with timber treads. We have used naturally processed materials which require no added decoration – wood, concrete and glass – and we are pleased with the resulting architecture.

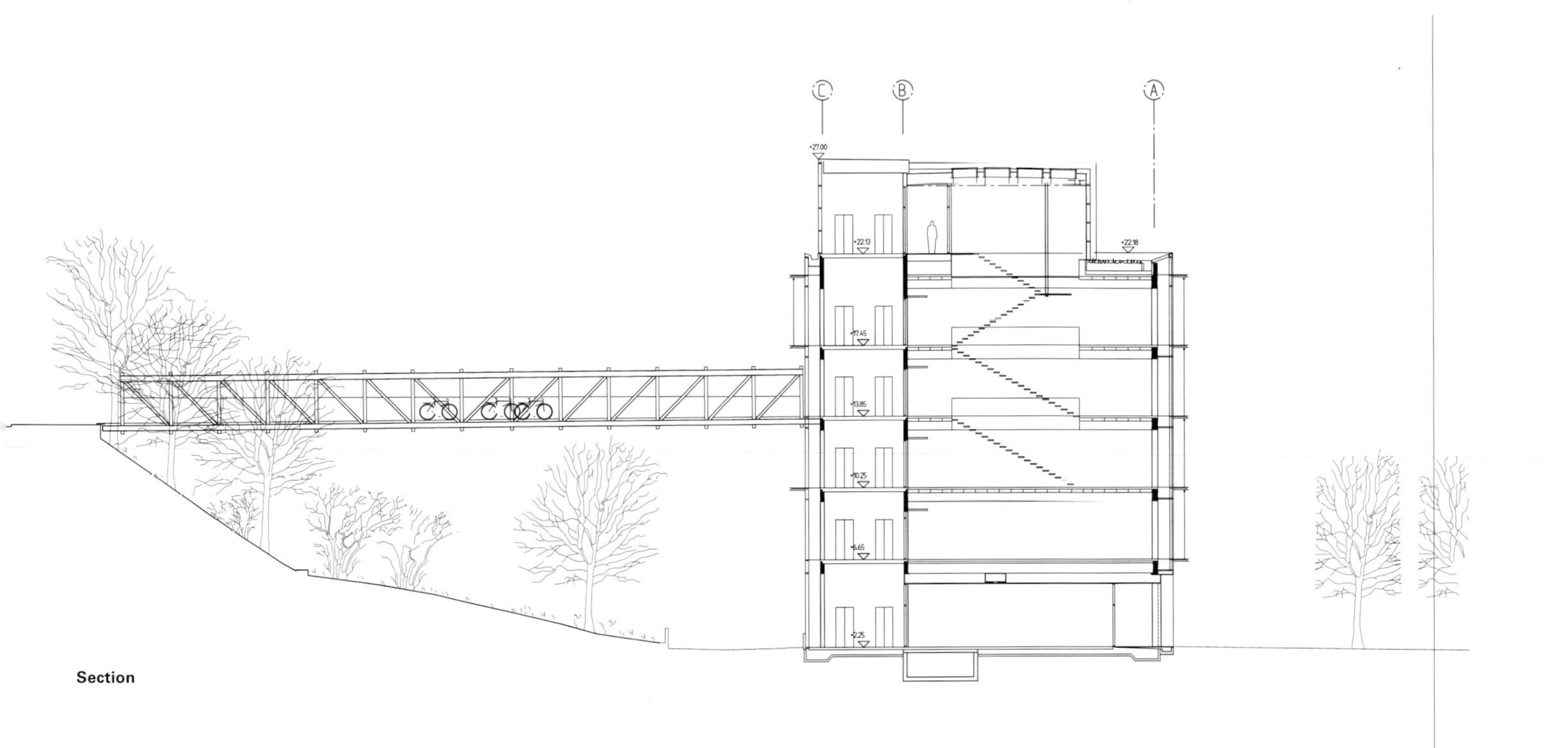

Section

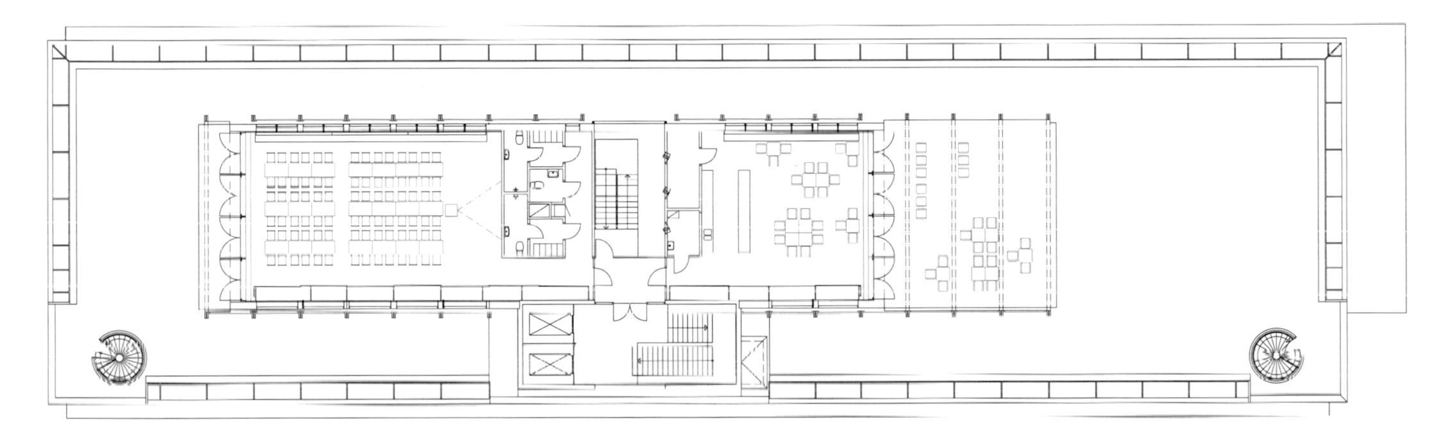

Fifth floor and roof level plan

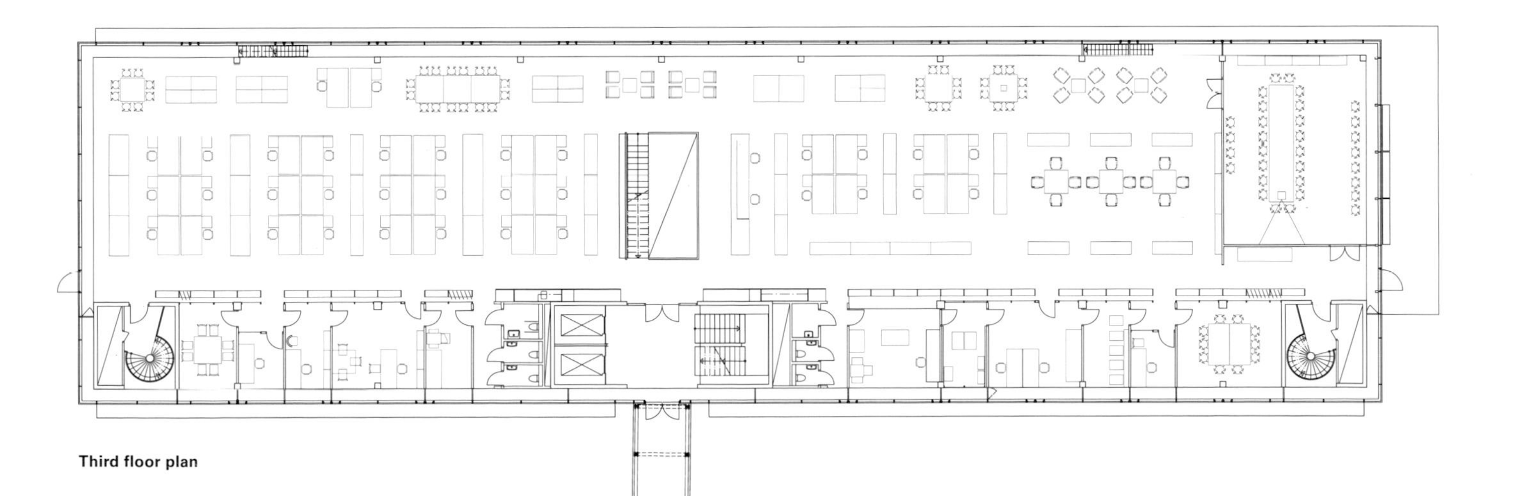

Third floor plan

Precast Construction

Arne Hellstrom, Strängbetong

The framework consists of concr
slabs. There is one floor below g
was built as a caisson, because
lake.

The three staircases with
of the building transferring the
Primary precast beams betweer
The concrete topping on the d
transfer the horizontal wind l
prestressed and of grade C60 c
concrete, carefully cast with as li
time of the framework was nin

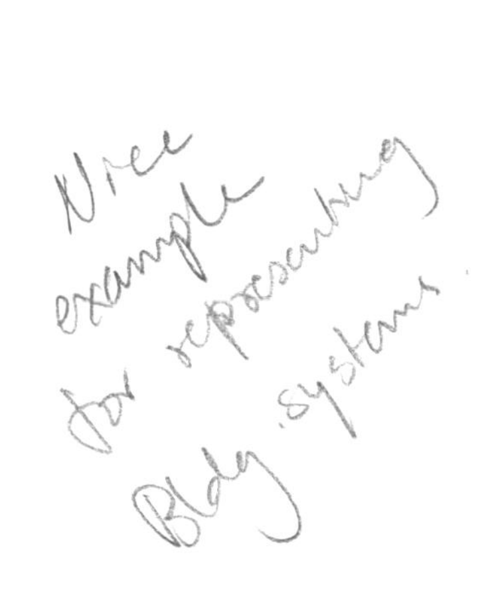

Installation of glass curtain wall

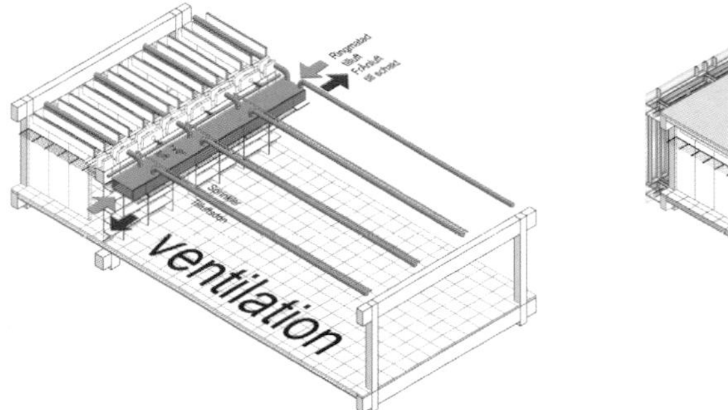

Ventilation system

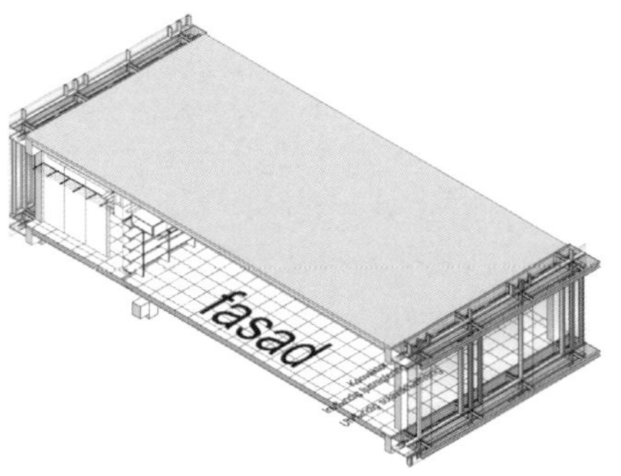

Façade detail

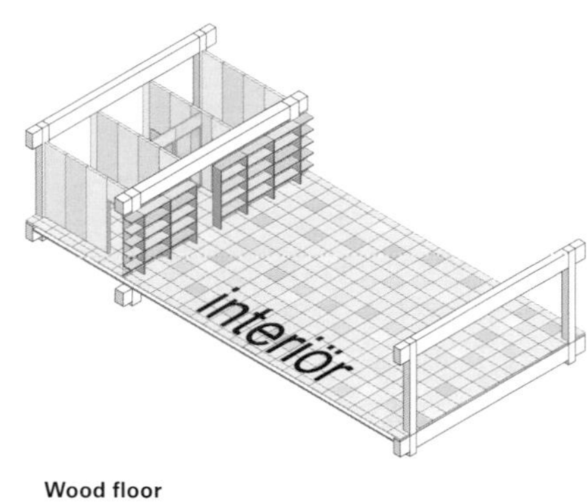

Wood floor and partitions

Main elevation

PROJECT DATA

Client and Architect: White Architects
Structural Engineer: Scandiakonsult
Mechanical Engineer: Angpanneforeningen
Electrical Engineer: Elkonsult Lennart Goldring
Contractor/Developer: PEAB
Precast Manufacturer: Strängbetong
Total Floor Area: 6,752 m^2 (gross area)
Energy Consumption: 120kWh/m^2 per year
Completed: 2003
Construction Cost: 102 million SEK

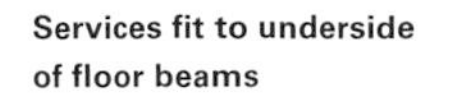

Services fit to underside of floor beams

View from waterfront

GEOLOGICAL AND GEOGRAPHIC SCIENCES BUILDING, STOCKHOLM

Nyréns Architecture

Location

The Geosciences Buildings are located within the Stockholm University campus area of Frescati. From the T-Centralen/Stockholm Centre you can take the metro towards Mörby Centre and get off at the stop University. From Odenplan you can go by Bus No. 70 towards the University.

Architectural Statement

The Geosciences Buildings gather several institutions previously dispersed all over the inner city. Two fan-shaped buildings that contain lecture halls and cafeteria are located in front of three tall regimented buildings that accommodate rooms for research.

There is a tradition and history with the use of concrete on the campus so it was important to provide the concrete connection. All five buildings are precast composite concrete framed structures. Precast elements are expressed on the façade of the rectangular blocks and the internal floor soffits. The triangular buildings were intended to have a board-marked concrete fascia, but after discussion with the client the architects opted for Swedish pine board panels, painted green.

Discussion

Anders Pyk, Dag Cavallius

The new buildings for the Geological Sciences department are in five blocks. The two for the undergraduate school are triangular in plan and the other three for the new research units are five storey rectangular blocks set out in parallel rows to the north of the student buildings.

The triangular buildings have green timbered façades and sloping roofs, they appear like huge pavilions in an expressive, romantic architectural style. There is no hint of concrete to be seen externally, but it is everywhere internally. The buildings contain a central lecture theatre and other smaller lecture rooms, in between there are a lot of light-filled public areas for a refectory, sitting area and a glass-panelled stairway that leads to the research building to the north. Internally the research buildings are like patrolled beehives with a central corridor on each floor leading to private research rooms on either side. Each is a closed celled building and this is reflected in the monumental, repeating bays of spandrels and columns that define the façade. It has a sturdy industrial feel to it born out of the grey precast elements that were selected for the envelope.

Garden and upper corridor

In a way the wood of the triangular buildings is a counterpoint to the massive, concrete dominated research blocks, and by introducing timber and a sloping roof the rigidity of the whole site is broken up. This was a logical response in our search for solutions that harmonised with the surrounding landscape, the green space in front of the building, and the oak trees we had to preserve in the garden courtyard. The triangular shape of the buildings evolved from the programme, the needs of the site and the widening scope of the brief. The open staircase and glass atrium that is the common access and communal corridor between the two blocks was designed as a timber-framed façade in natural wood, holding the large glass panels allow views into the courtyard garden and bring the oak trees and daylight into the building.

We visited a timber clad house designed by Frank Lloyd Wright in Chicago to explore our ideas. It was not the most celebrated of his buildings and not as well known because it was a traditional architecture with nothing particularly innovative, but we liked it very much. Having seen Oak Park House we sketched out our ideas and developed a layout for the triangular buildings, choosing to paint the pine slats. The concrete columns and wall elements were dictated by structural requirements. The floor has hollow core planks spanning between ribs of shallow T beams that were supported on deep precast beams. This is a typical composite construction system we use in Sweden.

Main stairway between the two groups of buildings

The building services were contained in the void between T beams and the soffit of the principal beam. In some areas the precast columns, particularly those near the open staircase that runs along the building perimeter, did not have a good surface finish so they were painted white.

The three research buildings to the north were arranged in a row with a repeating grid and identical external appearance. The column bays were at 6m intervals and they dominate the vertical line of the buildings by standing proud of the windows. They read as pilaster columns. On every other column there is an aluminium downpipe to carry rainwater from the roof and the cantilever ledges to the soak-aways in the ground. The windows are full height triple glazing units,

Front of timber-faced triangular building

Campus layout

Upper level research blocks

framed in timber. They sit between the upper and lower horizontal precast edge beams that span between the columns. Each bay window is made up of a series of discrete glazing units, some can open, others are fixed but with trickle ventilation slots.

At each floor level there is a continuous cantilever ledge emphasising the horizontal line of the building. It was not put there as an architectural feature but a practical solution to create an access to clean and maintain the window. It also screens direct sunlight and prevents rainwater from running down the building face.

There is transition between the taller triangular building and the rectangular block, and that demarcation we have clad in zinc. It is a bit like framing a picture with zinc overlaps along its edges. Precast concrete colour is a standard grey concrete supplied from the factory which suited our purposes. We believe in expressing the honesty of the materials, and prefer natural materials that weather well like copper, zinc, wood and grey concrete.

U-shaped precast column sits between spandrel edge beams

Top: Installing precast floor plank
Bottom: Edge beam with cantilever ledge being lifted onto column heads

Precast Construction
Arne Hellstrom, Strängbetong

The framework of the building mostly consists of precast concrete. The façades of the three rectangular blocks are of sandwich construction. The floor slabs are made of 200mm hollow core elements that are supported on the lower flange of shallow steel I beams that span between the precast columns.

Originally the building was designed as a steelwork frame with a precast façade. The inside of the single skin of precast would be insulated and there was to be a plasterboard lining and stud framing to the internal face. The concrete façade panels would act as edge columns and would be thickened to form horizontal beams over the top of the big window openings.

We then proposed that the whole frame was precast as it would be easier to construct and that the façade wall elements should be designed as sandwich spandrel units to eliminate the plasterboard and stud frame and the separate insulation layer. The façade elements were quite complicated, but less expensive and allowing for a better technical solution reducing cold bridging between the outer and inner wall panel.

We introduced the precast spandrel beam over the window head – it has a length of 6m, a depth of 1200mm and width of 600mm. It also carries the precast cantilever external ledge. The beam mould was built up so that the cantilever ledge was cast in with the beam. The inner face of the precast external column is U-shaped to allow for the fitting of service ducts. The width of the precast column was 900mm.

For both the horizontal and vertical precast elements on the external surface, we used a blend of 30% grey and 70% white cement. An air entraining admixture creating 5% air in the concrete for freeze-thaw resistance was added to the mix.

The structural framework for all three rectangular buildings was erected is six months.

PROJECT DATA

Client: Statens Fastighetsverk
Architect: Nyréns Arkitektkontor
Structural Engineer: Tyréns
Main Contractor: JM
Precast Manufacturer: Strängbetong
Completion: 1997

Undergraduate school

Exterior views of research building

Typical elevation of research building

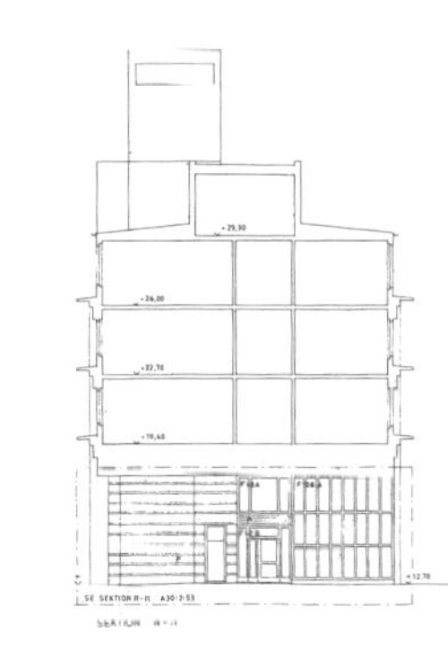

Sections

Foyer and waiting area to main lecture halls

STADTVILLA APARTMENTS, KASSEL

Alexander Reichel Architekten

Corner view

Location

The city on the River Fulda, located in the heart of Germany offers fast and easy access by train and motorway. The building is located in Unterneustadt, a new urban area of Kassel.

Architectural Statement

The modular principle of this town house is based on the brief for an architectural competition. The task was to design a building type for the eight different plots of this residential development on the outskirts of Kassel's Unterneustadt. Starting with a column grid of about 3m by 3.3m, this town house can be extended or modified to suit different uses and topographical conditions. One prototype was built as a straightforward cube measuring 13.52m by 12.3m by 15.4m; the other seven town houses were the responsibility of other prize winners. The building is set amid idyllic park-like surroundings not far from the River Fulda with its boat moorings and historic suspension bridge.

The use of full-height glazing to the living rooms allows the occupants to enjoy a view over the pleasant surroundings. The structure and the solid sections of the external walls are clad with glass fibre reinforced concrete panels; this artifice helps to indicate the different internal uses. The reinforced concrete frame members sometimes are clad in untreated larch wood infill panels. To emphasise the character of a detached villa, ancillary parking spaces are accommodated within the building itself by means of a mechanical car stacking system. A maisonette with a floor area of 120m^2 plus a low-level yard occupy the semi-basement and ground floor. The space can be used as an office or an apartment. The accommodation above can be divided to create two or three room apartments (plus kitchens and bathrooms). The top two floors are again maisonettes, and have a generous rooftop patio overlooking the river.

In order to achieve the desired variety in the façade and the necessary structural clarity, the building was divided into various systems; the load-bearing construction of reinforced concrete frame with precast concrete floors planks and walls, the timber framing elements and the cladding to the structural members. These individual systems are designed to remain visible in the façade and hence they organize the building's appearance. However, leaving a concrete structure exposed in Germany creates a building science problem. Owing to its high thermal conductivity, concrete must be insulated to prevent energy losses and damage caused by moisture. The concrete load-bearing structure was therefore clad with insulated precast elements. Glass fibre reinforced concrete (GRC) units just 30mm thick were chosen. In addition to their slim design and low weight, they are also easy to erect. The material and pattern of the joints of these accurate panels convey the structural rhythm of the concrete frame to the observer. GRC can be used as permanent formwork, as textured formwork or for rebuilding reliefs and cornices on older buildings, but here it is a façade panel. It consists of a fine aggregate concrete – the aggregate size is not more than 4mm – to which is added alkali-resistant glass fibre strands. These act as tension and anti-crack reinforcement. Each precast component is coated with a hydrophobic fluid at the works to produce a consistent, water repellent outer surface. This gives the surface a 'milky' shade which lends the material a vibrant quality.

The GRC panels were made by hand spraying both the concrete mix and the chopped fibres on to moulds. The unreinforced face mix of 5mm was colour matched to the grey concrete colour of the frame. This was coated with the reinforced backing mix that contained the chopped fibres whose length varied from 6-25mm and which were added at a dosage of 2% of the concrete volume. The backing mix was applied in five layers each of 5mm and compacted by rolling to bull up the panel thickness of 30mm.

PROJECT DATA

Architect: Alexander Reichel
Structural Engineer: Hoben, Kleinhans, Marx
Main Contractor: Hochtief AG
Completion: 1999

GRC panels and wooden shuttered windows

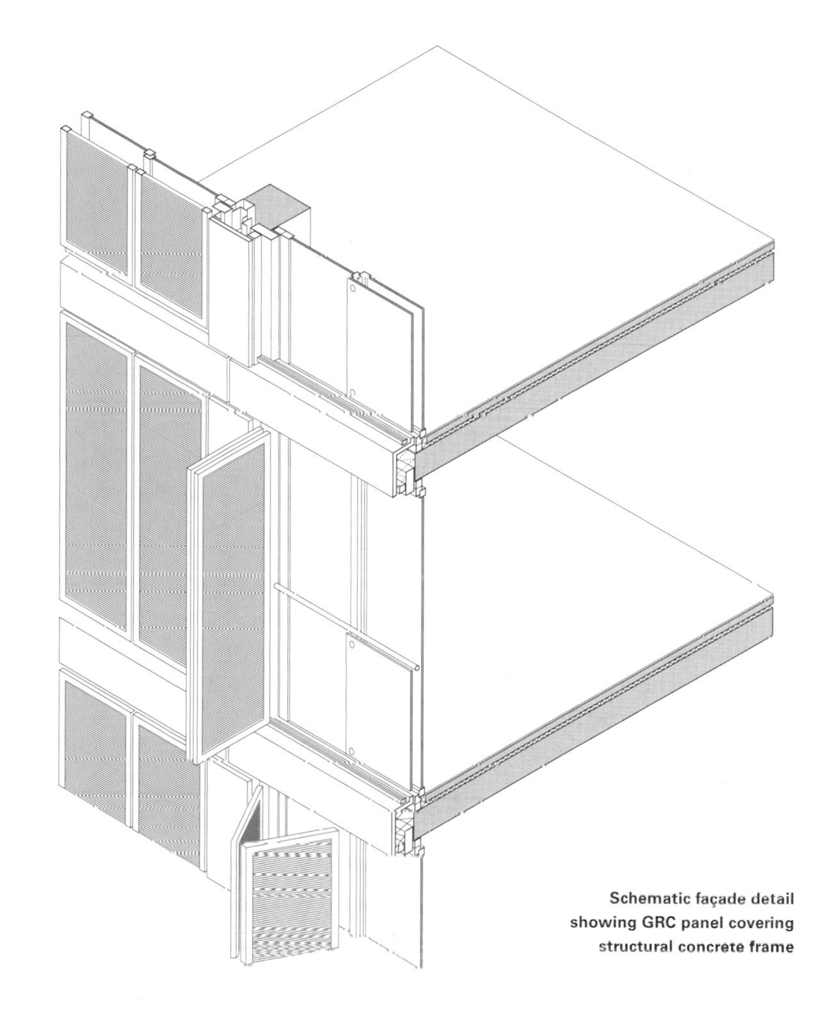
Schematic façade detail showing GRC panel covering structural concrete frame

Top floor terrace

25-35 PARK LANE, LONDON
Rolfe Judd Architecture

Location
The building is next door to the Hilton Hotel on Park Lane and Curzon Street and a short but pleasant walk from Hyde Park Corner tube station on the Piccadilly Line, in the West End of London.

Architectural Statement
The office development is formed around a newly landscaped square creating an attractive public open space and a tranquil approach to the main entrance from Curzon Street. This landmark building is made up of two new six-storey wings on either side of a five-storey listed building that has been fully modernised and refurbished. Built to a conceptual design by Michael Hopkins that has been reinterpreted and detailed by Rolfe Judd Architecture, the building offers 7,615m^2 of prestigious office accommodation with unrivalled views of Hyde Park.

With the flexibility of entrances from Park Lane or Curzon Street, the three blocks that comprise the development have basement car parking, lower ground and ground floor and six upper floors. It has been designed to meet the needs of modern day business, with air conditioned column free spaces throughout and interior finishes of very high quality.

The envelope is an aluminium framing system supporting Portland stone ashlar spandrels, double glazed bay windows, lightweight GRC column pilasters and GRC 'bird's mouth' reveals below the bay windows. Stainless steel brackets support the non-structural aluminium wind posts than run up the corners of the bay windows. At floor levels 5 and 6 there are brise soleils that overhang the façade and form a continuous balcony with a guard rail.

The structure is a reinforced concrete frame from basement to ground floor level and a steel frame with perforated steel beams and composite metal decking above it. The sixth floor steps back from the building line to accommodate the window cleaning track.

The cores for lift, stairwell and services are braced steel frames.

The receptions areas have green sandstone floors with feature up lighters and natural limestone walls with laminated etched glass partition walls. The doors are Canadian maple, the skirting is stainless steel and the suspended ceiling smooth white plaster.

Existing building and new wing

Architectural Discussion
Graham Fairley

Our brief was to work to the original concept by Michael Hopkins as they had planning approval for a generous office development, and revise the way the details worked so that the build cost would not exceed 17 million GBP. In essence we had to halve the façade cost without changing its appearance. The Portland stone, the precast and the cast aluminium had to remain but instead of designing the assembly as load-bearing with solid precast elements we designed the elevation as cladding, as a skin that was structurally redundant. The façade thus became much easier to build and faster to erect. Now we could build the structure and bolt on the cladding afterwards.

One of our first ideas was to make the solid precast column from lightweight GRC column shells which drastically cut down the deadweight while retaining the monumental appearance. Although we did not know much about GRC at the time, our later researches into the product and discussions with Trent Concrete and Techcrete, made it seem the obvious choice. The bay window trims and pilaster columns were designed as 20mm thick GRC units which would fit onto the curtain wall frame and be erected by the curtain wall contractor. The storey-high cladding units with the feature bay windows were designed as pods with a metal floor and steel edge beams that bolted to the structural floor. It had an aluminium roof with an aluminium steel trim that connected to the floor above. The glazing, the GRC spandrel and pilaster columns units and the flat Portland ashlar all had to be fitted to the cladding frame. The cladding package under the construction management contract was sent to a number of specialist curtain wall companies which Plus Wall won. As Techcrete were not interested in bidding for the GRC work it was tendered by BMS and Trent. GRC is a good architectural product with a fine surface finish but it is trivialised as a fanciful cheap product suitable for bird baths, Corinthian columns, hideous statues, Greek urns and artificial rock.

Trent won the GRC supply contract. We resolved the supply and installation of the cladding elements by having Plus Wall erect the whole system and Trent deliver the units for Plus Wall to fit on site. One of the spandrel moulds for the bay window feature was made slightly too large and we had problems of fit

Mould

Spraying GRC

GRC 'bird's mouth' reveals

Park Lane elevation

First floor plan

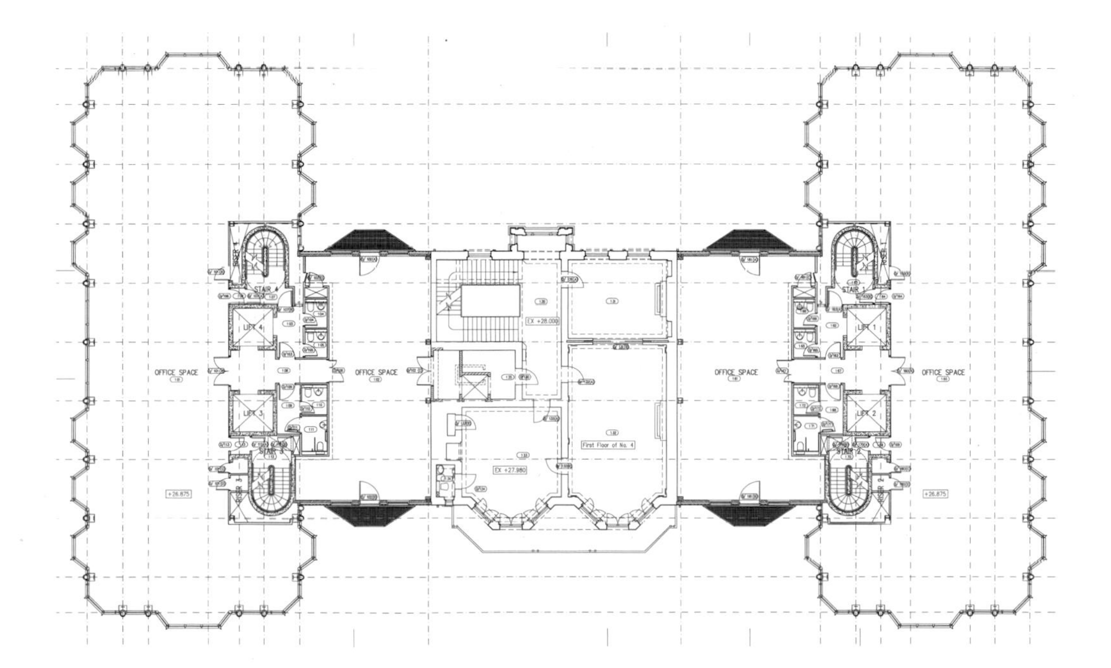

with the stainless steel node connections. There should have been a 20mm gap, instead the panel was 50mm too long. In the end Trent cut the end offs and repaired them. The mastic joint on one side of the panel was more than the other if you knew where to look. There was some crazing on the surface and a few arrises were not sharp.

The GRC finish and colour matched the Portland 'Grove Whitbed' ashlar supplied by Albion Stone and in many ways it was superior. We detailed the fascias to ensure that water was dispersed away from the panels and ashlar to minimise dirt staining. Where we have rain running on to the GRC spandrel panel from the glazing above it, we have sloped the panels inwards from top to bottom and introduced a small drip detail. There is metal backing to the panel which is the effective water barrier and protection to the insulation. I always remember studying buildings that Lutyens designed and noticing the way he sloped the stonework to reduce water marks and dirt staining.

GRC Construction

David Walker, Trent Concrete

We worked closely with the architects and the façade engineer to detail the fixing arrangement and check the panel integrity for movement and rigidity.

The panels were sprayed into ply mould lined with GRP. First the facing mix coat of 4mm colour matched to the Portland Stone sample was sprayed on. Then the backing coat with the alkali resistant glass fibre strands was sprayed again in 4mm layers and rolled for compaction, to build up the 16mm backing thickness. The surface was given an acid-etched finish and then supplied to Plus Wall on site to install them under our supervision.

The GRC pilasters and bay window reveals combine with flat ashlar panels

Bay windows keep rainwater away from GRC

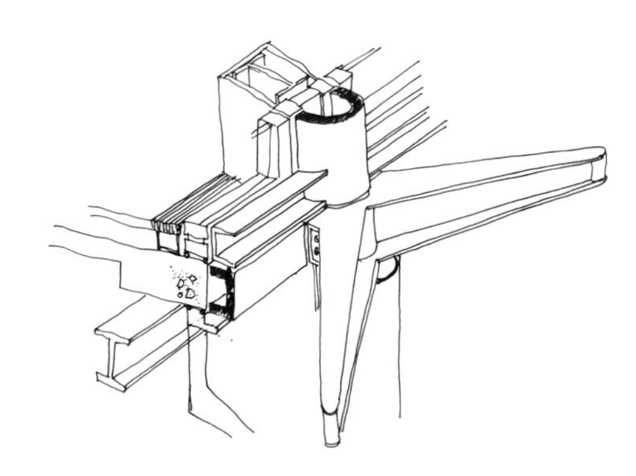

Detail of cladding joint

PROJECT DATA

Architect: Rolfe Judd Architecture
Structural and Services Engineer: Buro Happold
Façade Engineer: Arup Façades
Development Manager: Taylor Warren Developments
Construction Manager: Heery International
Curtain Walls: Plus Wall
GRC Manufacturer: Trent Concrete Ltd
Completion: 2002
Contract Duration: 2 years

Park Lane elevation

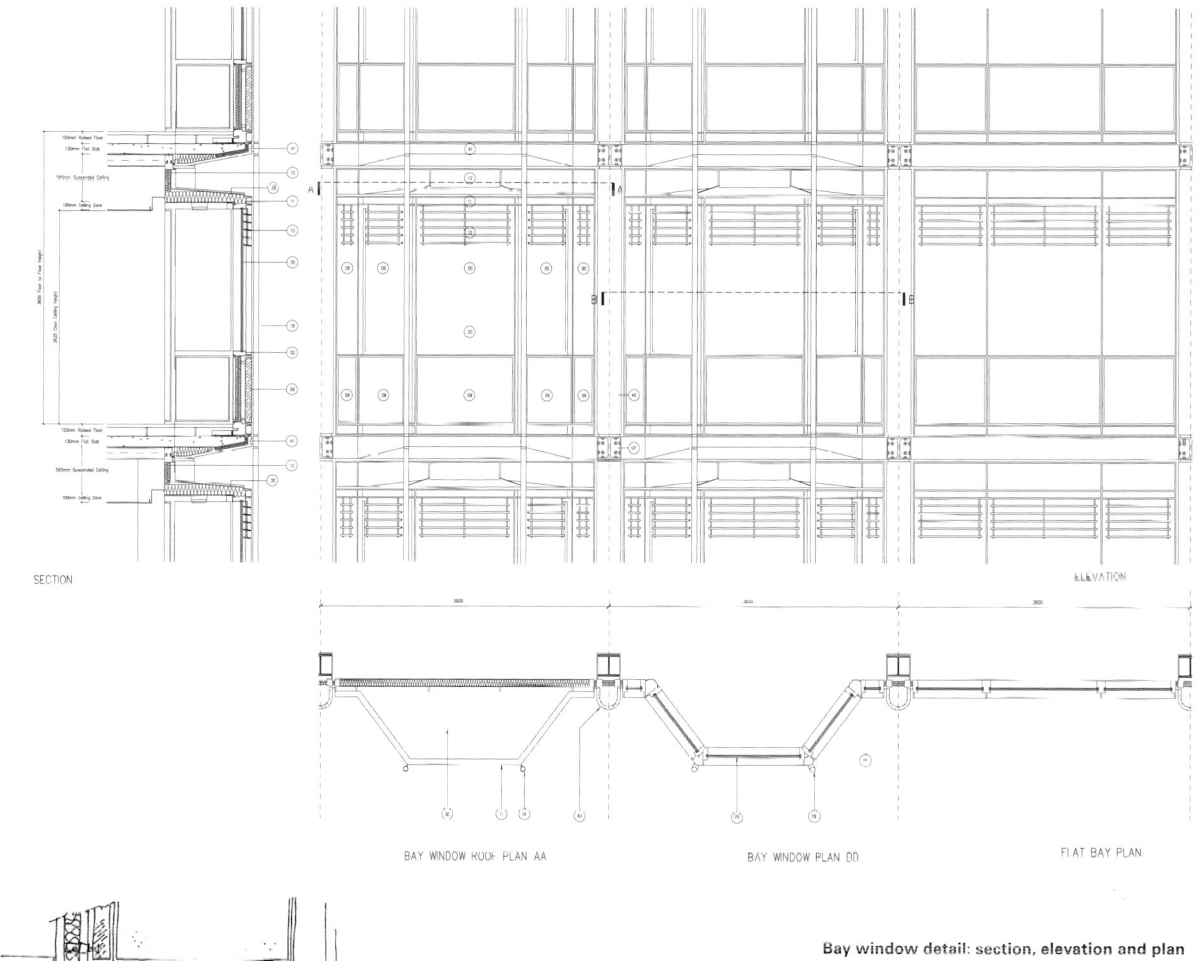

Bay window detail: section, elevation and plan

Cantilever overhang of bay window at floor level

GRC pilaster columns

SPIRAL STAIRCASE, COPENHAGEN
Arkitema Architects

Location
The staircase is located in Tuborg 15 building which is along Tuborg Boulevard in the new urban quarter of Tuborg Havn in Hellerup. The area was a run-down industrial zone only a few years ago, since then it has changed into a vibrant new commercial district of Copenhagen.

Concept sketch of exterior

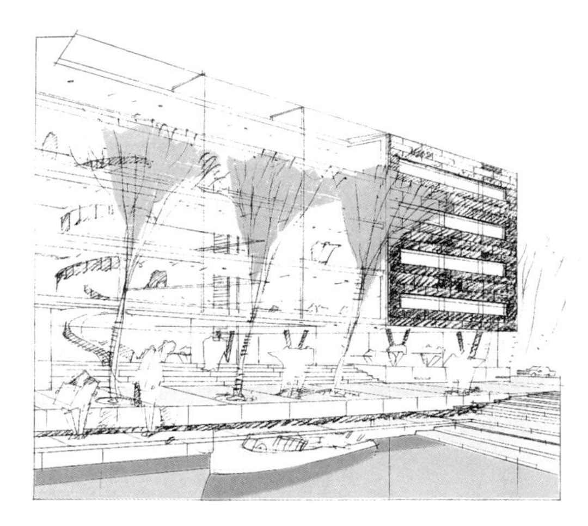

Architectural Statement
Rolf Kjaer, Arkitema Architects
Tubor 15 is a purpose-built four-storey office building that is leased by three software companies. The staff share facilities within the building such as the foyer and the restaurant located in a bright spacious atrium where daylight floods through the glass canopied roof. Large trees and a water fountain give the atrium a touch of the great outdoors. The open staircase spiralling the east elevation of the atrium is a distinctive feature, a piece of sculptural art. Connected to the floor divisions at landing level only, it is a completely self-supporting structure.

When we first thought about the spiral staircase we had never heard of CRC, so we intended to scheme it in steel or reinforced concrete. It was only by chance that I came across CRC which is a fantastic product. We met with Bendt Aarup from CRC Technology and with our engineering colleagues at Ramboll and thus the concept of the spiral staircase evolved.

Our first idea was to design the staircase with a double balustrade and no columns and all in white concrete. Our engineers persuaded us that it was better structurally to have a column and only an interior balustrade. As we preferred not to have a detached square column, we shaped the column so it formed an integral part of the balustrade and was curved. The proportions of the spiral were then modelled on a computer using a 3D imaging programme. The leading edge was made as thin as possible as it was an important focal point. CRC is grey in colour due to the grey micro silica even though it is made with white cement. So we had it painted white using a silicate paint commonly applied to exposed concrete in Denmark.

You will see that there is a wooden plinth made of ash at the base of the staircase. We wanted to make a transition from the stone floor of the atrium to the staircase and this was the solution. After many discussions with glass manufacturers we found one company that could make a curved laminated glass for the outer balustrade. The glass panels are supported at intervals by metal upstands bolted through purpose-made holes on the edge of the CRC tread. CRC is very hard and tough and it was easy to bolt on the upstand connections. The handrail of the glass balustrade is a specially turned and curved piece of laminated ash which fits precisely over the top of the glass.

The stair treads and risers were faced with ash, to muffle noise from metal tipped shoes and to reduce the hardness of the surface. We introduced optical fibre lights along the inner balustrade just above the step line.

Handrail with laminated ash

Main elevation, Tuborg 15 building

Structural Considerations
Hans Exner, Ramboll Consultants
The column support was necessary, although the architect had sketched the design concept without columns, merely suggesting parallel balustrades that connected to the building at floor level. The precast beams at the edge of the building floor were both too weak and too thin to carry the landing loads from the staircase when we assessed the structure. We could neither fix nor tie the balustrade walls to the building floor without having to do some strengthening work to the entire edge of the floor beam. This would have been disruptive, costly and quite unsightly. It should be noted that the staircase was included after the main building had been designed and building floors had been constructed. We also felt that the double balustrade wall would not look as pleasing as it blanks out the transparency and lightness of the staircase.

We proposed a column with an interior balustrade wall acting as a beam from which the steps cantilever out. The leading edge of the steps can be made very thin as there was no force acting on it. However the column was made as a curved segment of a circle, wider at the base and tapering towards the top and

Concept sketch of atrium and staircase

View from staircase

Cantilever treads and curved column profile

an integral part of the balustrade. It supports the spiral beam and the landing and carries the loads to the foundations. The balustrade beam spans from floor to floor, between the column. The column restricts the bending and torsion in the balustrade beam. Each staircase step cantilevers as an independent element – we did not allow for any interaction or restraint from the adjacent steps nor for the spread of load in our calculations. In reality the steps and risers help each other and ensure an exceptional rigidity of the construction, resulting in zero deflection at the edge under the worst loading conditions. We designed the floor loading as 2.5kN/m^2. The landings are connected to the building floor and provide the lateral stability of the staircase structure.

Every element of the spiral staircase is precast with 100Mpa CRC. The flights – which comprise the balustrade beam and cantilever steps – are cast in four sections and stitched together with in situ JointCast CRC on site. The joints between flight are nominally 80mm wide as the JointCast CRC has an exceptional high bond strength which reduces the joint width. The upper balustrade section connects over the top of the column and links with the landing section. The first flight structure is carefully propped before the upper column element is put in position and the joints then filled.

The main difference between CRC and conventional concrete apart from the increase in compressive and tensile strength, is the superior bond strength and anchorage length that we can work to. We can design with very short lengths of starter and lapping bars, only one fifth of the bond length required for normal concrete. The most critical section was the anchorage length of 100mm required for the rebar of the cantilever steps. Allowing for tolerance and cover this required the balustrade wall to be 150mm wide. The reinforcement takes all the cantilever moment and transfers it to the balustrade beam. Perhaps we could have made the balustrade beam as thin as 100mm and used U bars to develop the anchorage for the steps, but that would have made the joint details with the interconnecting bars of the balustrade beam too complicated. In any case we felt we needed the 150mm thickness to cater for the bending and torsion in the beam.

When we first proposed a column supporting the balustrade beam we suggested it should be square. The architect came up with the idea of making it 150mm thick the same as the balustrade and to curve its width to maintain the curvature of the balustrade. That was a very elegant solution which we then detailed.

In all our design work using CRC we had to justify our calculations and assumptions to the checking authorities. We showed them the long-term test results on durability, bending, anchorage, fatigue etc. that CRC Technology had undertaken over 15 years. As regards fire risk, the staircase is not the fire escape stairs for the building and therefore required only a half hour fire rating. The concrete cover to the bars was 15mm (10mm for cover and 5mm tolerance) and we have used 8mm diameter bars in the steps and 16mm bars in the balustrade wall and column.

Construction of spiral staircase

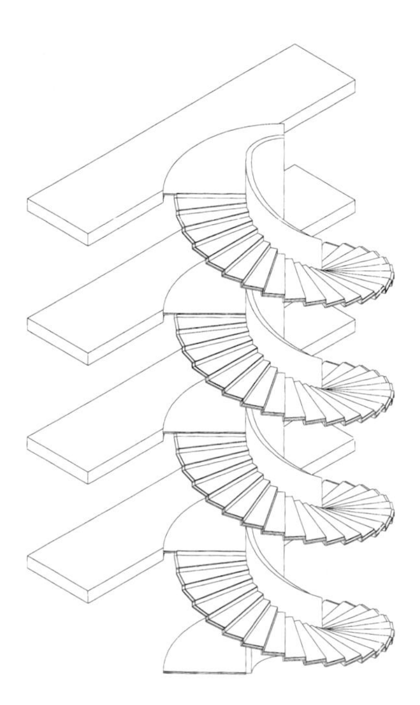

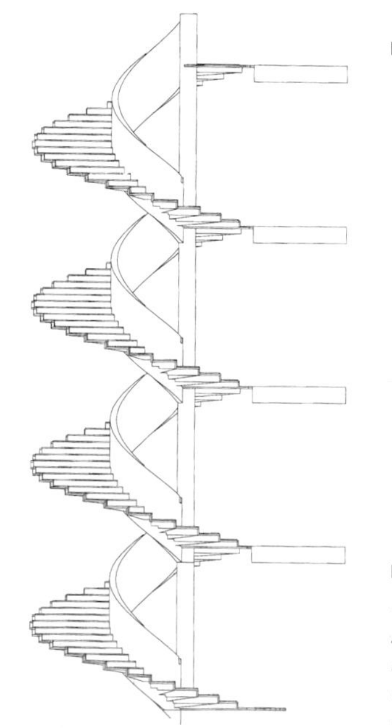

Final design schematic

PROJECT DATA

Client: Carlsberg Ejendomme
Architect: Arkitema
Engineer: Ramboll
Contractor: NCC Denmark A/S
Completion: 2002

Looking down the spiralling balustrade

Atrium and spiral staircase

Concrete sculptural art

SEONYU FOOTBRIDGE, SEOUL

Rudy Ricciotti, Bandol Architects

Location

The Seonyu footbridge links the Sunyudo Island on the Han River with the city of Seoul. The island has been inaccessible to the public since 1976 but now that it is converted into a beautiful park for the public to enjoy, the footbridge provides the access.

Architectural Concept

Rudy Ricciotti

In Seoul's broad cityscape considered on the scale of conurbation, structures spanning the river are numerous, while others are still under construction. There are bridges for car traffic whose construction is based on steel or concrete technology. The steel bridges make use of heavy trussed girder construction of 100 years ago. The concrete structures express bridge technology of the 1960s with concrete piers and beam and box girder deck sections. In this context it seemed natural to adopt a more audacious technology.

The scale of the span is the most difficult aspect, for the distance of 120m is too short to make reference to the car bridge structure that brushes the top of the island. It is also too great a span to consider the work from the landscape perspective alone. What is required then is a proposal deftly combining sign and meaning. The first sketches put forward by the city of Seoul showed a suspension bridge in the style of the Golden Gate, breaking both with the scale of the site and the use of modern technology and materials in spanning the river.

It had thus to be a concept that would not rupture the sight line and the narrative of the landscape. The main motif was a taut arch with the slimmest depth over the 120m span.

Its exceptionally sleek proportions are to evoke the smoothness and white colour of porcelain and fragility of an egg shell.

Design and Construction

Mouloud Behloul, LafargeDuctal, France

The Seonyu footbridge, which was built in time for the World Cup in 2002, was called 'The Bridge of Peace' when it was opened but has since reverted to its official name. It consists of two steel approach spans and a central arch of 120m made of Ductal®.

Ductal®-FM, with a compressive strength of 180Mpa, was specified for the Seonyu Bridge where high bending and direct tensile strengths are required. These mechanical properties are achieved by introducing short steel fibres 13-15mm in length with a diameter of 0.2mm at a dosage of 2% of the mix volume. The application of heat treatment after the mix sets in the mould, eliminates drying shrinkage and greatly reduces creep.

The Ductal® Arch

The 120m arch is connected at each end to massive reinforced concrete foundations which are 9m deep. These foundations are designed to absorb the horizontal thrust of the arch. The arch consists of a ribbed upper deck slab (the walkway) and two girder beams in a double T configuration. The width of deck slab is 4.3m and the beams are 1.3m deep. The deck slab has a 30mm topping with transverse ribs at 1,225mm centres. The depth of the ribs was 160mm for the first and sixth segment and 10mm for the others. The deck slab is supported by the two 160mm thick girder beams. The shape of the girder beams and deck slab geometry was chosen for easy demoulding of the section.

The ribs of the deck slab are prestressed by either one or two 12.5mm diameter monostrands. Specially adapted small anchors were used to transfer the prestressing forces from the strands to the ribs. Each girder beam is prestressed longitudinally by three tendon clusters which are sleeved through metal ducts. There are nine strands in each of the clusters in the lower two ducts and 12 strands in the upper duct. The tendons of the beams are stressed once the segments are in place on the supporting scaffold towers. After completion of the stressing phase, the tendon ducts are grouted. Two temporary monostrands are cast into each segment in the lower part of beam to cater for stresses during lifting and

Aerial view

Properties (typical values) of Ductal® with steel fibres and after heat treatment

Density	2,500kg/m^3
Compressive strength	180 MPa
Tensile strength	8MPa
Post-peak strength in tension	5MPa
Young modulus	50,000 MPa
Poisson ratio	0.2
Shrinkage	0
Creep factor	0.2
Thermal expansion coefficient	12.10-6m/m

placing operations as each segment in positioned onto the scaffold towers that were built across the river.

The arch is composed of six segments. These segments are prefabricated in an area next to the final location of the arch. Diaphragms are added at the ends of each segment. The diaphragms on the end segment spreads the compressive loads impacting on the foundation concrete, while those over the central arch are for jacking the two halves of the arch.

The segments are 20-22m long and curved. The slope at the extremities is more than 8%. The volume of Ductal® in a segment is 22.5m^3. The total mixing time to fill the metal mould for each segment was 5 1/2 hours. The mould is filled using eight injection points positioned midway along the internal surface of the beams. During the casting operations, the fluidity of the mix is constantly checked and controlled.

After casting a segment, it is cured in the mould at 35° C for 48 hours. A spreader beam is used to crane lift the segment from the casting area to a heat treatment chamber. The segment is then steam cured at 90° C for 48 hours.

The six segments (three on each half of the arch) are positioned in sequence on the scaffold towers by a crane, mounted on a river barge. The segments on each of the half spans are stitched together, then prestressed before the tendon ducts are grouted up. The two half spans are finally joined together by casting the short in situ crown or key segment stitch. Before casting the in situ stitch a

Mould design and mould structure

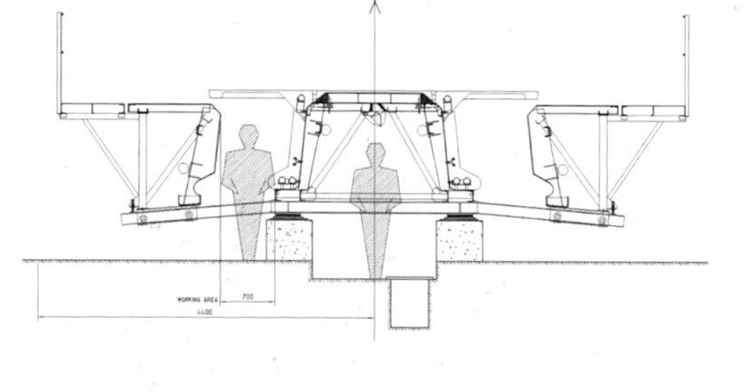

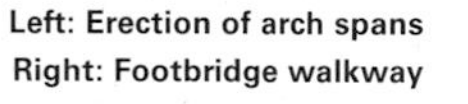

Left: Erection of arch spans
Right: Footbridge walkway

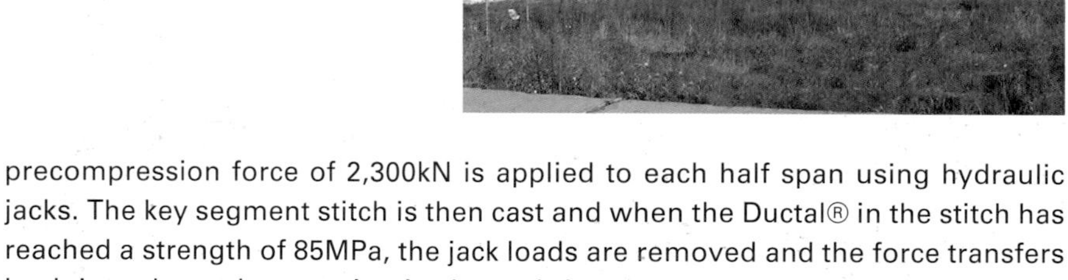

precompression force of 2,300kN is applied to each half span using hydraulic jacks. The key segment stitch is then cast and when the Ductal® in the stitch has reached a strength of 85MPa, the jack loads are removed and the force transfers back into the arch, to maintain the arch in precompression. This is good for stability and robustness.

Potential vibration of such a slender arch had to be considered. Analysis produced an elegant solution based on shock absorbing tuned mass dampers which limit the horizontal acceleration to 0.2m/sec and the vertical acceleration to 0.5m/sec to maintain crossing comfort level.

This is the first time in the world that an ultra high performance concrete, reinforced with steel fibres has been used for a span of 120m. The properties of Ductal® have made it possible to design a very slender arch with thin sections, giving the footbridge elegance and grace.

PROJECT TEAM

Architect: Rudy Ricciotti, Bandol Architects
Contractor and Engineers: Bouygues Travaux Public
Prestressing Work: VSL (Korea)
Lighting Design: Yann Kersale
Completion: 2002
Bridge Dimensions: Length 120m;
height (at mid-span above water level) 15m;
depth of section 1.3m; deck slab 30mm;
width of deck 4.3m